THE TRIUMPH OF TECHNIQUE

OTHER WORKS BY ROBERT WOLF

Books

Crazeology: The Jazz Life of Bud Freeman
Story Jazz: A History of Chicago Jazz Styles
Heartland Portrait: Stories and Essays of Rural Life
An American Mosaic: Prose and Poetry by Everyday Folk
Jump Start: How to Write from Everyday Life
The Triumph of Technique: The Industrialization of Agriculture and the Destruction of Rural America
Violence in the Promised Land: Witnessing the Conflict in the Middle East

Plays

Lucrezia: A History That Never Happened
The Austringer
Edward the Confessor
The Special Prosecutor (with Wayne Julin)
The Arms Dealer
The Diplomat
The Strike at Pullman
Heartland Portrait

THE TRIUMPH OF TECHNIQUE

The Industrialization of Agriculture
and
The Destruction of Rural America

Robert Wolf

Art by Bonnie Koloc

Ruskin Press
Waukon, Iowa

 First edition. ISBN 0-9741826-0-5. Printed in the United States of America,

1. American agriculture—19th and 20th centuries.
2. Technology. 3. Medieval commerce.

RUSKIN PRESS
P.O. Box 66
Waukon, Iowa 52172

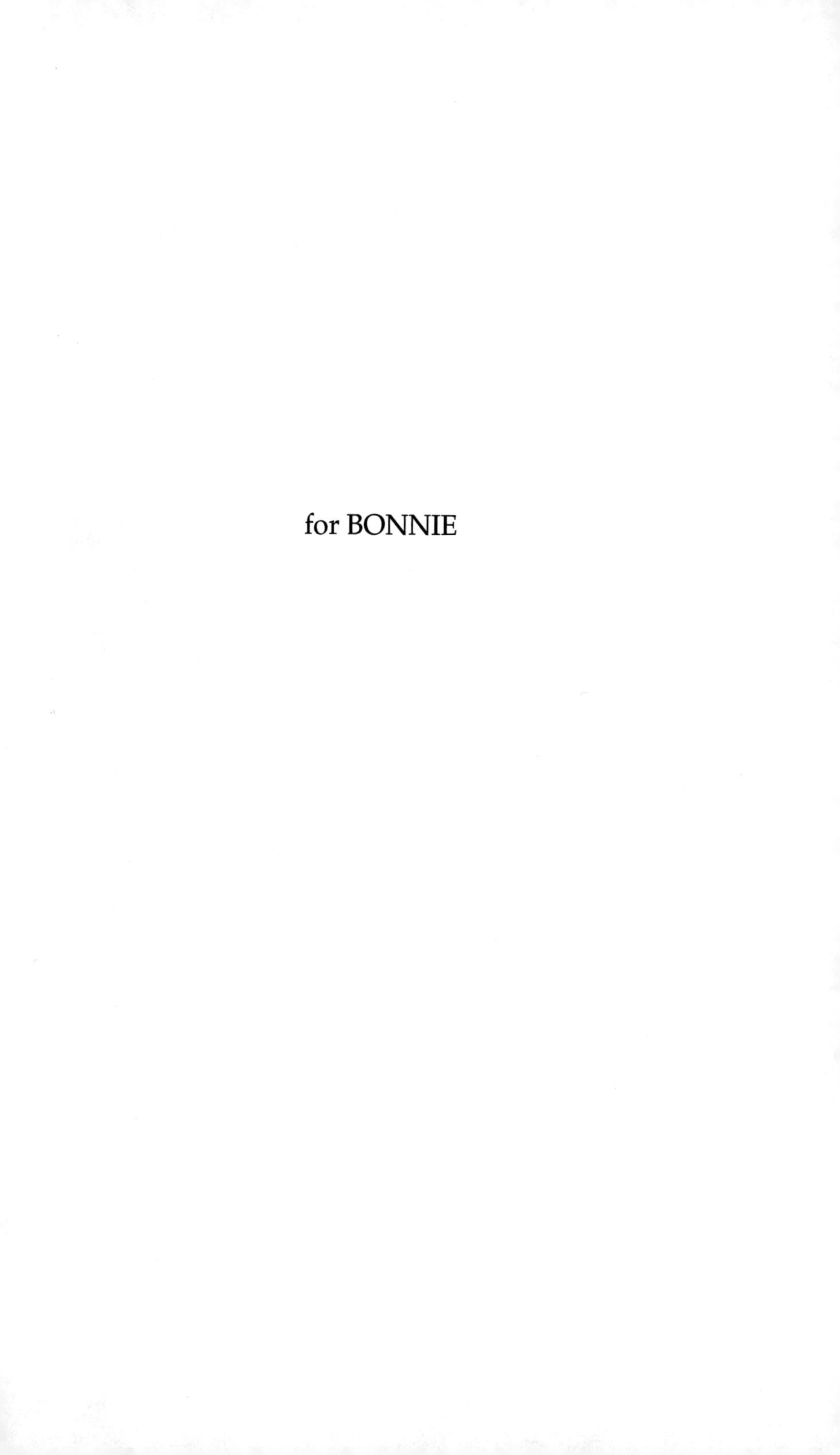

for BONNIE

WHERE TILLAGE BEGINS, OTHER ARTS FOLLOW. THE FARMERS THEREFORE, ARE THE FOUNDERS OF HUMAN CIVILIZATION. Daniel Webster

TABLE OF CONTENTS

PREFACE

Argument

Technique is an idea. Techniques are its offspring, tools and methods for accomplishing predetermined ends. Over the past five centuries, as we have sought ever increased efficiency in all the practical arts and sciences, technique has developed into an autonomous entity, determining every facet of contemporary life.

In agriculture no less than the other practical arts, technique has played a determining role. Besides providing the instruments for the creation of a centralized economy, technique has spawned numerous tools, such as pesticides and chemical fertilizers, bioengineered crops, and confined animal feeding operations. These and other technologies have transformed agriculture into agribusiness, thereby taking the art out of farming.

Large scale operations dominate the rural landscape. Gone are the small farmers who could earn a living off their land. Now medium size farmers face extinction. As human scale farming disappears, so too do rural towns, and thus the death of rural America coincides with the triumph of technique.

Acknowledgments

Earlier versions of material from this book appeared in other publications. Chapters four, five, and six are expansions of three articles originally written for *The North American Review*. Chapter two is an expansion of an essay written for *More Voices from the Land*, published by Free River Press and reprinted in *An American Mosaic*, published by Oxford University Press. Chapter three is a revision of a lecture delivered at the University of Nebraska, Kearney for the Distinguished Lecture Series.

The publication of *The Triumph of Technique* was made possible primarily through the generosity of the Bradshaw-

Knight Foundation and the Jesse Smith Noyes Foundation, with additional help from Iowa Citizens for Community Improvement.

I am indebted to Dennis Keeney, professor emeritus at Iowa State University and senior fellow at the Institute for Agriculture and Trade Policy, for his extensive notes and comments on the first draft. Thanks are also due Christopher Bamford, editor-in-chief of Lindisfarne Books, for advice which helped transform the book from a series of essays into a unified work. Professor Jon Andelson, Director of the Center for Prairie Studies at Grinnell College, made very useful comments on a later draft and gave it a test run in his course, "Nature and Culture on the Prairie." Fred Kirschenmann, Director of the Aldo Leopold Center for Sustainable Agriculture at Iowa State University, pointed out the need for a discussion on industrial organic and helped fill the gap. Neil Hamilton, Professor of Agricultural Law at Drake University, was generous in critiquing chapter six, although there are points at which we disagree. Finally, for their input and encouragement I want to thank Jay Knight, President of the Bradshaw-Knight Foundation; Robert Karp and Practical Farmers of Iowa; Don and Mary Klauke, Rural Life Directors for the Dubuque Archdiocese; Murray Hudson, writer, antique map dealer, and activist; and Paul Becker, Publishing Director of Iowa State Press.

INTRODUCTION

The Triumph of Technique grew in part out of my longtime desire to know America, a desire that led to years of roaming the country, working at dozens of jobs. For years it meant hitchhiking and riding freights trains back and forth across the continent, or driving hundreds and sometimes thousands of miles to look for work or to see old friends. All of this began in adolescence with a passion to discover the American soul, which meant having to know every area of the country, to live in every town and city, to work every job.

My experiences kept multiplying as every few years I moved to a new state or town and invented a new living for myself. In 1991 my wife and I moved to northeast Iowa. There, living on a farm ten miles from the nearest town, I immersed myself for the first time in the rural Midwest and learned the precarious nature of the farm economy. I learned that the farm crisis of the 1980s was not over. We saw it in the farm auction notices posted weekly on cafe bulletin boards and in hardware stores, and in the numbers and statistics tabulating the actual decline published after each USDA farm census. As I continued to read about the farm crisis, I began to see how the federal government had betrayed the farmer, even the small banker, and in general had been hostile to the rural village and economy. Today rural Americans harbor few illusions about their economic future. They have seen small farmers continue to leave agriculture and they see the consequences on Main Street, where small retailers struggle to survive with fewer customers and against the encroachments of large national chain stores.

Soon after moving to rural Iowa I learned about the existence of militias in our area, and I came to understand that the federal government's role in the demise of rural society provided impetus to their growth. I had been told by a group of Minnesota farmers not far from us that one of their neighbors had been arrested by the FBI for having what the Minnesotans described as "enough dynamite to blow up the entire valley." Across the Mississippi River in La Crosse County, Wisconsin, a group of men had declared the court system illegal and had issued arrest warrants for the district attorney and others.

Eventually the harsh realities of the farm economy confronted my wife and me. As declining farm prices continued driving small farmers out of agriculture, some saw industrial agriculture as their only hope. One of our neighbors, a hog farmer whose father had also raised pigs, was faced with the choice of leaving farming or growing hogs on contract for Murphy Family Farms. He chose the latter, and while we understood the tensions underlying his decision, my wife and I joined a lawsuit to stop his operation, which had the potential to pollute the area groundwater, lower our property values, and pollute our air. It was then we learned just how the cards are stacked against the individual: those of us opposed to the operation naively assumed that the Iowa Department of Natural Resources was on our side. But the employees of the DNR area field office to whom we complained literally laughed at us and obstructed every move we made. They were political appointees of a governor who had worked hard to woo hog factories to our state.

It was then I understood how powerless we were, and this sense of powerlessness evoked enormous anger. Since the politicians had brought the hog confinement operations to Iowa, they could care less how these operations affected rural residents. Obviously the bureaucrats—political appointees—cared as little. Their goal, in fact, was to thwart our attempts to obtain justice. The more I read, the more I saw how far corruption had spread throughout the agricultural system, nationwide. I could understand how militias are formed; I could understand how violence is the last recourse of a people stripped of legal protection. It was my anger, born of my powerlessness, that led me to write this book, hoping in some way, to whatever slight degree, that it might help to turn the tables on those businessmen, politicians, bureaucrats, and others who were working against the regeneration of agrarian society.

2.

Shortly after I moved to Iowa, I heard former Senator Warren Rudman claim that if the United States did not eliminate its deficit that it would become the world's largest banana republic. I then began thinking about economic collapse, which led me to wonder just how large an area might increase its self-sufficiency to the extent that it could avoid much of the trauma of a national depression. One town could not do much on its own, but northeast Iowa could at least supply its own food, provided it had a bartering system in place. Southeast Minnesota and southwest Wisconsin share the same topography as northeast Iowa—rolling hills and valleys built on a karst foundation. I began to think of these three adjacent areas as part of one region. An acquaintance told me that it had a name—the Driftless region—and that it included a small sliver of northwest Illinois. I began to think that by virtue of their shared topography, and the fact that all four agricultural areas share a common culture, that the region's residents might be persuaded to work cooperatively on economic development.

Without question the Driftless region could grow all the food it needed. Our ancestors worldwide had done as much, and we, even with our more diverse and ample diet, could do the same. In Spain and Russia, for example, multitudes of greenhouses grow vegetables in colder months. One relatively small area in Spain contains scores of greenhouses that raise vegetables sold across Europe. The Sloga Project in Macedonia, a closed agricultural system, raises sheep, cattle, and goats whose waste is the basis for a humus that is nourishment for trout, mushrooms, pear and apple trees. The trees are pollinated by bees that produce honey. In sum, this system makes cheese and sells it along with milk, meat, honey, trout, mushrooms, pears and apples.

Such a system, with modifications to allow for local diets, could be adopted anywhere. The Chinese have a closed system that in addition to producing food, creates electricity from vegetable and animal wastes. Bio-gas stoves and bio-disgesters fueled with these wastes typically provide a Chinese commune with up to two fifths of its electricity. Closed agricultural systems on the scale of the Macedonian and Chinese projects—one of China's closed systems feeds 90,000 people—could be built in the United States and would satisfy a growing demand for a decentralized agricultural system of local production for local consumption.

To minimize costs, such a system would have its own plants for processing meat, milk, and vegetables, and would be considerably smaller than our current centralized plants. In the case of meat processing plants, a local plant would be far less likely to sell contaminated products because its smaller scale would allow for slower production. Considering the fact that our centralized system of production and distribution has resulted in each of the states importing approximately 85 percent of its food, a decentralized agricultural system begins making sense for several reasons, the most obvious of which are lower fuel consumption and the restoration or building of a local farm economy.

Local and regional self-reliance could be fostered in other areas of the economy. Community development banks devoted to financing low income housing and small business start-ups, along with local currencies (formalized barter systems) are two tools for creating credit and wealth. Imagination is all that is required to see what possibilities exist, while courage and a willingness to cooperate with other towns and individuals would realize these possibilities. Heaven did not mandate a centralized economy of scale; let us remember that the United States was economically and culturally decentralized until the late nineteenth century. The present system's comforts and conveniences, combined with our apparent power over Nature, has led people to believe that contemporary culture represents a high point in human development. That the converse might be true is a proposition that few of our political and business leaders have ever pondered.

3.

The Triumph of Technique began as a series of articles on the industrialization of agriculture but quickly expanded its scope beyond what is usually considered in a survey of the subject. As industrialized agriculture is not simply one aberration in an otherwise healthy culture, but one of numerous manifestations of malaise, a fuller understanding of it necessitates looking at its several roots.

The most obvious was the development of industrial production in the nineteenth century and the subsequent development of the railroad, which made national distribution possible. Together the two created a national economy which not only enabled farmers to buy agricultural equipment from distant cities but made it possible for middlemen to sell crops and processed foods nationwide. But not only was Midwestern wheat transformed into bread and sold in Chicago, New York, and Savannah, not only was Mississippi cotton woven into cloth in Massachusetts mills and then transformed into clothes sold across the nation, but by the late nineteenth century American grains and cotton were flooding European markets too. This transformation of farming in the nineteenth century from an art for immediate use into a science of production for profit depended upon a more significant change: the psychological transformation of the farmer himself.

But neither machine production nor national distribution nor the transformation of the farmer would have been possible without another, far less obvious, agent of change: the merchant class. The merchant now dominates all social and economic activity worldwide, but his ascent from a position of disrepute and suspicion to one of admiration and honor cannot be fully understood until one understands a more fundamental shift in Western history, a shift in Western man's apprehension of what is real. For this reason the third chapter of this essay deals in part with the fourteenth century philosophical battle which ended with the defeat of idealism and paved the way for empiricism, a change of outlook which has had momentous consequences for the West.

4.

Agriculture's industrialization raises two immediate concerns. The first is the potential health risk posed by most food grown and processed in the United States, including genetically engineered plants and animals; the second is the ongoing destruction of the family farm and the consequent demise of rural America. The intimate connection between small towns and the family farm is such that as small-to-medium-size farms continue to fail, the rural economy weakens and fewer Main Street businesses thrive. The lure of the big city has long beckoned to rural Americans, and the exodus from the farm began well before World War One.

But the demise of the family farm and small town is due not only to the lure of cities and lucrative jobs, it is due in large measure to the mechanization of agriculture, or as it is now called, the industrialization of agriculture. This mechanization includes not only the widespread use of tractors and motorized combines and other diesel-driven equipment, but the use of chemical poisons for weed and insect eradication, and most recently, the genetic engineering of crops and animals. As more and more mechanization has been introduced into agriculture, costs have continued to skyrocket until high input costs combined with continued low returns for crops and livestock have conspired to drive more and more family farmers from the land.

The USDA's census of agriculture shows that in 1997 the number of U.S. farms was at its lowest since 1850. Between 1992 and 1997, the number of farms earning less than $100,000 per year fell from 1,591,437 to 1,565,671, a decrease of 25,766. When one eliminates all farmers earning less than $10,000 from the 1992 and 1997 statistics (which will eliminate many of the hobby farmers), one finds that the number earning between $10,000 and $99,999 decreased by 82,015. Small farmers, as these figures show, are on the decrease and the trend will undoubtedly continue, considering that the average age of farmers continues to rise. In 1954, 15 percent of all farmers were under thirty-five, while in 1997 the number fell to 8 percent. At the same time, the number of farmers fifty-five or older grew from 37 to 61 percent. As of this writing there is little doubt in the minds of rural Americans that the family farm, as a widespread institution, will never be revived. Some small- to medium-sized operations have prospered, but they are largely organic farms. The number of these sustainable farms, however, remains small in comparison to the total number of farms.

The fact that small farmers were persuaded to accept labor saving devices is understandable. And one can also understand how the apparent efficiency and seeming profits to be derived from chemicals and bioengineering convinced farmers to adopt both. That is, one can follow the reasoning that persuaded them to convert to these techniques. But farmers should have seen that with each new technique they adopted there was a further diminution of their numbers.

The process of industrializing agriculture indicated that some basic shift in American sensibility and thinking was manifesting itself. Once people moved off the farms they eventually forgot the smell and feel of earth, the smells of animals, the sounds of storms, the feel of the land awakening in spring or frozen under winter snows. Amid tall buildings few saw the sunrise and sunset, and with ever-present artificial lighting they could no longer see the stars. The living processes of the earth became, increasingly, an abstraction. The presence of death had been with them on the land when they slaughtered a steer or lamb or hog; but when, in the nineteenth century, packaged meats appeared, there came a generation that had never seen a living hog or steer, and whose flesh now became a mere thing, abstracted from the living animal.

The farmer eventually became almost as alienated from Nature as the city dweller: sitting atop his tractor he lost the intimate connection that the plowman had following his team: his feet no longer felt the ground, and his hands, arms, back, and legs no longer responded to the yank and pull of horse or ox. The more technologies that were introduced, the deeper his alienation grew. Work so laborious and time consuming that it once required teams of men, could now be completed in solitude. Before the days of agricultural science, the nurturer had to wrest his living from the land; now the earth was made to yield its harvests through the efforts of men in faraway laboratories and design offices. The efficiencies kept multiplying: fertilizers, hybrid crops, insecticides, herbicides, bioengineered crops and animals.

A farmer utilizing all these efficiencies might think he is connected to the land, but the fact that he has placed numerous, highly sophisticated technologies—heavy equipment, chemical poisons, genetically modified seeds, perhaps a mechanical feeding system—between himself and untended Nature, argues that we are dealing with a technician, not an artist or nurturer, and that his experience is far removed from that of his forebears. His is a state of mind that sees Nature as a field for unlimited interventions, which means that whatever operation he might perform with seeds, plants, and instruments is legitimate.

The fact that agriculture is now thoroughly industrialized means that the urban sensibility, which is to say the sensibility which sees Nature as a thing, has, for practical purposes, definitively triumphed over the organic world view. Nature is no longer living, it exists. Despite the talk of some biologists about ecology and field and interrelatedness, the majority of biologists perceive Nature in precisely the same terms as the man or woman raised among high-rises or amid miles of suburban development. Nature consists of things and parts, and as such can be manipulated howsoever they feel or think it should be manipulated.

And so this book is also about the death of Nature, which had to precede the industrialization of agriculture, for the latter could not have developed until we were persuaded that Nature was a mechanical system, for only then could we subject her to mechanical operations. The significance of the adoption of the mechanical world view cannot be overstressed: the consequences are seen about us in the gradual destruction of the biosphere through global warming, ozone depletion, the desertification of the land and the destruction of the ocean, and most recently in the approaching genetic engineering of human beings. Whether through global warming or man's genetic engineering, the result is the same: the obliteration of human life. To think, as many contemporaries are wont to think, that an organic paradigm of nature is now in the process of supplanting the mechanical world view is hopelessly out of touch with the data. It shows more completely perhaps than anything else contemporary man's incapacity for an unsentimental assessment of affairs.

As the first chapter shows, some current technologies are known to produce deleterious effects; others are thought to do so. All have been widely discussed within the farm community, with the result that there is no farmer who can plead ignorance of known consequences. Some farmers, for example, realize the dangers of chemicals and have reduced their use; many others are indifferent to their effects on the environment or people and continue heavy application. For them farming has become just another business, and they do whatever they must to maximize profit and remain in operation. How can such a one claim to feel connected to Nature?

The technician's apology for extreme intervention, including genetic manipulation, lies in the claim that he is merely working with Nature to improve her efficiency. On the surface, his argument resembles that of the ancient Greek physician Galen, whose work *On the Usefulness of the Parts of the Body*, frequently reiterates that Nature is an artist and that medicine achieves its effects by imitating and guiding her operations. The technician's goal of altering and attempting to control every variable affecting a plant's growth—the soil, nutrients, and seed—is to create a portable environment and plant that will enable agribusiness to grow the same crop anywhere on earth. What motivates the technician is in fact control over Nature and the construction of artificial substitutes for organic substances. Far from considering such a project to be working with Nature, it is in fact an attempt to supplant her. It is an act of unsurpassed arrogance and hubris that will soon run its course.

Our teachers with respect to hubris are the Greeks, for whom pride came before a fall. Their tragic poets saw the lesson embodied in myths and wrote it out in epics and tragic dramas; two of their historians, Herodotus and Thucydides, saw it in the rise and fall of states. It was a moral law, a truth of human nature, that success without self-knowledge leads to blinding pride, which leads inexorably to collapse and ruin. The only antidote to pride is embodied in another Greek saying, the Delphic injunction: Know thyself. But this means nothing to the modernist, who is a rationalist for whom limits do not exist, whose motto well might be: What we have accomplished today is nothing compared to what we shall accomplish tomorrow. For the modernist, technology can create or transform virtually anything to match our needs. Eventually, even old age and death need not be feared.

Such a mentality claims the entire world for its province but is crippled with a memory that does not recall yesterday's failures and a foresight that cannot consider consequences for the next generation. Coupled with the drive for profits, this mentality inevitably creates problems from which our civilization will be unable to extricate itself. Its collapse, like that of all other civilizations, is inevitable, but can only be hastened by its blind reliance upon technology, including agricultural techniques.

5.

The final chapter of this book focuses on technique and its social and environmental consequences. Technique, which encompasses not only technology but any rational method designed to attain predetermined results, has absorbed every facet of our lives. Technique not only designs industrial and military products, but management flow charts, public relations and advertising campaigns, livestock production systems, pharmaceuticals, psychotherapies, educational curricula

and instructional methods, and almost very other f
culture and economy.

Ancient cultures had techniques, but they w
by virtue of the fact that they were still a means and not an end in themselves. Today the commitment to efficiency and uniformity and profit dictates the continuous development of new techniques, and once the endless proliferation of techniques became a mandate, science was reduced to a handmaid of industry. In the seventeenth century, however, science was still independent of technique, and while Bacon and Descartes had each claimed that scientific advancement would make us "the lords and possessors of nature," the work of Newton, Huygens, Galileo, and their contemporaries was still undertaken in the spirit of disinterested inquiry.

In *The Technological Society*, Jacques Ellul makes the point that European treatises on non-scientific subjects in the sixteenth and seventeenth centuries were a grab bag of knowledge: a scholar discoursing on one subject would display his knowledge in numerous areas, and an essay on law or medicine, for example, might include history, comparative religion, metaphysics, botany, metallurgy, and anything else the writer may have studied. But beginning in the eighteenth century, writing became increasingly dominated by technique, which meant that scientific and legal writings narrowed their focus. While the main concern of the earlier writer, according to Ellul, was to make a full presentation of himself, the later writer has been concerned to impart knowledge for use. With the advantage of hindsight we can see that the narrowing of focus has entailed a general and significant loss of understanding of the interconnection of things. Thus contemporary scientific technique requires specialists so highly trained that not only is it difficult for them to stay abreast of current literature in their fields, it becomes impossible for any to have more than a passing understanding of discoveries in other highly technical areas. Even if a researcher were concerned with the impact of his work on society and the environment, discovering those consequences could entail years of costly research.

Now that science is primarily concerned with accumulating facts in the service of industry, most scientists are now essentially technicians. A focus upon facts, as opposed to principles and hypotheses, means that technicians understand their subjects in a very limited sense. They can easily fail to understand how the object of their study behaves under other sets of conditions, with other variables, and the impact that it will have on anything outside their preselected environment.

We find numerous examples of how the technicians' limited understanding, when applied to agriculture, leads to health and environmental problems. To take but one example, antibiotics in animal feed greatly increases rate of growth and weight in livestock, but since antibiotics are transferable to humans, this technology has contributed to the spread of antibiotic resistant bacteria and become a contributing factor in a growing health crisis. An understanding of the interrelationship of phenomena would have caused preliminary studies on the transferability of antibiotics. But western man's predisposition to exploit any technique for immediate technical advantage, especially when it has the potential for generating profits, now threatens our very existence.

6.

The present essay considers industrial agriculture from multiple perspectives, and given the brevity of its style, the reader could easily forget or lose key points. The capsule manner of presentation has encouraged me to adopt a symphonic form in which the first chapter is equivalent to a musical overture, announcing the various themes to be enlarged upon in subsequent chapters. Those themes, like musical motifs, are repeated with variations throughout the book.

CHAPTER ONE
OVERVIEW

The matter of contemporary agriculture is as good a way as any to penetrate the depths of the modern world, into the nihilism that lies at the heart of the intersection of contemporary business, politics, technology, science, and law. Agribusiness, as our highly mechanized agriculture has come to be called, is what its name implies: the transformation of agriculture into a form of business or commerce. Its proponents—university scientists and economists, politicians and businessmen—seek to displace what remains of the world's indigenous agricultural production with an agriculture of scale, claiming that without highly sophisticated methods of production and distribution we will be unable to feed the next billion people in the world's exploding population.

Far from achieving that goal, agribusiness's efficiency demands fencerow-to-fencerow cultivation, the development of monocultures, the use of large machinery, and the application of insecticides and herbicides—all of which contribute to soil degradation and topsoil loss. Fencerow-to-fencerow cultivation entails the destruction of hedgerows, which protect against wind erosion. The reliance upon monocultures—the production of one crop on the same field year after year—necessitates the use of chemical fertilizers, insecticides, and herbicides. These, besides adding dangerous pollutants to surface and ground water, kill the life in the soil, most significantly earthworms and bacteria, which supply humus and maintain soil structure. With earthworms and bacteria absent, soil is more easily compacted, leading to more water runoff and consequent topsoil loss. Compaction is further aggravated by the use of heavy machinery.

Business, however, can only make enormous profits

from agriculture when production, processing, and distribution are highly mechanized, centralized, and in the hands of relatively few companies. To insure that its program is carried out, the seed companies, processors, implement manufacturers, and other agribusinesses make enormous research grants to schools of agriculture to get the imprimatur of university scientists and economists; in addition, they make large contributions to political campaigns in exchange for favorable state and federal legislation.

But very few people outside of rural America know the extent to which potentially lethal techniques dominate contemporary agriculture, and of the failure of courts and politicians to protect the public from the rapacious corporations that control them. Major newspapers seldom cover agricultural issues, and when they do, often provide inadequate accounts. This unfortunately comes at a time when the world's food supply is seriously compromised, even threatened. For the moment, let us briefly consider five areas of immediate concern as a way of entering into more detailed consideration of the situation.

First, a handful of corporations—thanks to mergers, acquisitions, and invisible holdings—are now running the overwhelming percentage of farm and farm related business in this country. They and the press call it agribusiness, and rightly so, because methods of corporate farming and food processing aim solely at profits, regardless of costs to the environment or people. Vertical integration in agriculture—which entails a cluster of allied corporations owning or controlling businesses involved in the production, processing, packing, and distribution of a particular commodity—is the secret to corporate success. Vertically integrated corporations have a virtual stranglehold on agriculture, offering the family farmer what agribusiness is willing to pay for crops and livestock, with the result that family farms continue failing at a dizzying rate. The danger that this high concentration imposes on consumers is that we can expect the quality and safety of food to become ever lower as these corporations squeeze every last dime of profit from their operations.

Second, the genetic engineering of plants and animals is proceeding at a staggering pace, in laboratory outposts in the farthest reaches of the world, run by corporations that are patenting the rights to their novel species. Tens of thousands of technicians are employed in decoding the DNA sequences of plants and animals, and genetically engineering what amount to new species, crossbreeds between two disparate entities. Many prominent scientists have warned of the dangers that this technology presents to the entire biosphere, especially in cases when the mode of transport across species involves pathogenic (albeit disabled) viruses.

That anyone would think they had the right to take that which emanated from Nature or the Divine, depending on your point of view, and mix it in an almost arbitrary fashion, ignoring millions of years of evolutionary history, is an act of hubris which is staggering but which, apparently, leaves thousands of businessmen, politicians, scientists, technicians, and farmers without a qualm. The corporate motive is clear, and that the United States Supreme Court sanctioned this absurdity is incomprehensible, until one realizes that a society based on philosophical materialism has no basis in law to oppose it.

Third, most livestock, including chickens, hogs, and dairy cattle are now grown in large sheds where they spend all of their lives penned until slaughter, and in such conditions that large amounts of antibiotics are mixed with their feed to help increase their weight and cope with disease incurred by stress. Hogs grown in confinement and cattle finished in feed lots are also given steroids for weight gain. Both antibiotics and growth hormones are transferable to humans, with negative consequences. Growth hormones in meat result in weight gain for the eater, while the overuse of antibiotics has resulted in the development of strains of antibiotic resistant bacteria and a consequent major health crisis.

Fourth, pesticides and chemical fertilizers remain an enormous environmental problem, the most recent chemical devastation occurring in the Gulf of Mexico, where a dead zone has been created by chemical runoff from farmlands draining into the Mississippi River. Extensive corn production is responsible for this. It takes 1.2 to 1.3 pounds of nitrogen to develop one bushel of corn, and the Midwest's intensive corn production, which has continued for well over half a century, has called for enormous quantities of nitrogen laden fertilizer. After a heavy rainfall this nitrogen leaches out of the soil into rivers and streams or moves below root level, into ground water. In the Midwest, the nitrogen that has leached into creeks and rivers eventually drains into the Mississippi and flows down to the Gulf. Within that large body of water off the Louisiana coast, nothing lives. Nitrogen causes hypoxia—oxygen deprivation—which has killed slowly moving bottom feeders and forced faster moving species farther off shore.

Fifth, politicians pretend concern for the small farmer, the consumer, and the environment, but owe their political life to the very corporations that are putting the small farmer out of business and are now poised to destroy a portion of the biosphere. Thus, even when a majority of their constituents oppose hog confinement operations, U.S. senators from farm states refuse to speak against them directly, but instead address questions concerning manure disposal. The fact that manure storage lagoons at hog confinement operations have poisoned water supplies, most conspicuously in North Carolina, has not prodded the U.S. Congress into action.

The fact that politicians are seemingly unwilling to oppose corporate interests means that their politically appointed bureaucrats in the federal and state agencies charged with enforcing the law often ignore or deny environmental damage and potentially life-threatening situations. In a USDA memo obtained in 1999 by *Mother Jones News*, the department admitted it was under pressure "from above" to approve genetically engineered foods, which it did. Making matters worse, this bias in favor of large corporations extends to politically appointed judges who, for the most part, side with them. Finally, the enormous concentration of agricultural power in relatively few hands has been given the blessing of the anti-trust division of the U.S. Department of Justice, which has consistently approved agribusiness mergers and buyouts.

2.

From the non-farming public's point of view, the most radical consequences of industrial agriculture are its hazards to health and the possibility of high food prices. For years farmers have said that the era of cheap food is about over, and they may be right. Once the three or four corporations that are now creating new species of plants and animals for commercial use have completed their assignment, those Americans who cannot grow their own food will have no choice but to pay whatever the corporations will charge for what they choose to grow.

As for the consequences of genetically engineering our food supply, we have had warnings from numerous scientists, among them Nobel laureate and biologist George Wald, who pointed out that because cancer causing viruses are used in genetic engineering, the world may find itself swept with new diseases, even plagues. Let us remember the problems in the nuclear, chemical, and transportation industries which led to incidents at Chernobyl, Love Canal and the Galapagos Islands—to name just a few of our most conspicuous environmental calamities. Let us also remember that DDT was going to kill mosquitoes and lice, and thus reduce malaria and typhus, without harmful side effects.

Perhaps most important, let us remember that agricultural chemicals, along with their industrial counterparts, were supposed to help usher in a technological Golden Age, bringing prosperity and well-being to all, but instead have been linked to increased risk of cancer. In his note "Pesticides and Cancer," Dr. Rob Rutledge quotes a National Cancer Institute finding that "farmers tend to have higher than expected rates of cancer of the lymph, blood, lip, stomach, skin, prostate, testes and soft tissue . . ." and that "phenoxy herbicides, triazines, arsenicals and organochlorides play a role." A 1999 Swedish study conducted by doctors Lennart Hardell and Mikael Eriksson, established links between glyphosate, a

Monsanto chemical marketed as Roundup (the world's most widely used pesticide), and non-Hodgkin's lymphoma.

Very possibly more significant than either of these is the presence of dioxin. Dioxin is the family name for chlorinated chemicals which are manufactured for use in pesticides but are far more often generated by incinerators when chlorine compounds are burned with hydrocarbons. While 95 percent of all dioxins are released into the environment through incineration, lesser amounts are produced in paper making, steel manufacture, metal recycling, and lumber milling. Dioxin released through incineration is carried by wind virtually everywhere and settles on plants, which are ingested by animals. When it settles in the water, it is absorbed by fish. The EPA estimates 25 to 30 million pounds of 2,4-D (which contains dioxin) is used as an agricultural pesticide. 2,4-D was one of the components of Agent Orange, the defoliant used by the U.S. military in Vietnam. Twenty-five to 30 million pounds of 2,4-D may seem small, but experiments have shown that even minute traces of dioxin can cause hormone disruption. Add this to the dioxin falling onto plants and into streams, lakes, and oceans and the problem is considerable.

The most comprehensive study of the dioxin problem is found in *Our Stolen Future*, by Colburn, Dumanoski, and Myers. Colburn and her colleagues correlated hundreds of studies of the effects of dioxin on animals, which demonstrate that dioxin disrupts hormone functioning throughout the endocrine system, including the thyroid, pituitary, pineal, and reproductive glands, to produce a variety of diseases and abnormalities.

The body mistakes hormone disrupters such as dioxin and DDT for hormones. These synthetic chemicals are picked up by receptors, which are proteins located within cells targeted by the hormones. In normal functioning, hormone and receptor bind and move into the cell's nucleus where they can activate genes to create proteins that can affect growth or sexual characteristics, or change the activity of existing proteins to alter the blood sugar level or heart rate. Some disruptors block hormone functioning; others mimic it.

Many scientists are now convinced that dioxins are a cause of a host of problems in the human population, including breast, prostate, and testicular cancers, as well as lower sperm count, smaller male genitals, decreased female fertility, and ovarian dysfunction that are surfacing worldwide. This conclusion is suggested by three facts: one, that dioxins are known to cause reproductive failure in animals; two, estradiol, an estrogen present in both men and women, is identical in animals and humans; and three, that the hormone receptors in all species are strikingly similar. In other words, if dioxin can fool hormone receptors in animals into treating it as estradiol, it may do so in humans, given the similarity of animal and human receptors. Thus it would appear that a hormone disruptor causing a testicular malformation in alligators or other species would cause it in humans too. The fact that malformations and cancers of the reproductive system have appeared over the last twenty years with increasing frequency in humans suggests that dioxin is responsible. If so, then our future is in jeopardy.

In agriculture, at first glance, efficiencies appear to have produced nothing short of miracles, but to measure success simply in terms of crop yield is illusory, for the real measure of an agricultural system's success or failure is in its long term impact on a population's well being and its own sustainability. By that standard our agricultural system has failed us dismally. Chemicals have poisoned our food, soil, and water, while the totality of agricultural efficiencies continue to help drive farmers from the land, thereby undermining the future of rural communities. And in the long term, our farming methods, if continued indefinitely, will deplete and (as we will see) acidify our remaining top soil.

3.

By the year 2000, the crisis of American agriculture had passed, for by then not even the most sanguine observers believed that the family farm or rural village would regain the prosperity and vitality they had enjoyed fifty to a 100 years before. The year 2000, as an approximation, marked a moment as decisive for rural America as 1880 had for the frontier. That latter year, according to the Superintendent of the Census, in a report issued in 1890, marked the frontier's closing. In Frederick Jackson Turner's famous analysis, the frontier had shaped American institutions and character, and had been the continued spark for American democracy. In "The Significance of the Frontier in American History," Turner wrote: "The peculiarity of American institutions is, the fact that they have been compelled to adapt themselves to the changes of an expanding people—to the changes involved in crossing a continent, in winning a wilderness, developing at each area of this progress out of the primitive economic and political conditions of the frontier into the complexity of city life." Elsewhere in the essay he noted: "American social development has been continually beginning over again on the frontier." The American character, he said, had been shaped by this continual rebirth from primitive conditions, this struggle with Nature. At the close of the essay, Turner announced: "And now, four centuries from the discovery of America, at the end of a hundred years of life under the Constitution, the frontier is gone, and with its going has closed the first period of American history."

Throughout the history of the frontier, the human presence had been marked by a change in tools and techniques of ever-growing sophistication. Urbanization and mechanization moved with the frontier, following behind, but always following. The frontier's closing marked a swing of the balance away from wildness and agrarian civilization toward urbanization and mechanization. Indeed, after 1880 the industrialization that had exploded with the Civil War further accelerated, and the last two decades of the nineteenth century marked the appearance of laissez-faire capitalism and industrial empires, and a neo-Darwinist ethic to match.

As mechanization and urbanization grew, so did centralization. If 1880 marks, approximately, the crisis in the clash of industrial and agrarian forces—the point at which the forces of centralization and urbanization overcame the countervailing tendencies towards decentralization and agrarian existence— then the year 2000 marks, also approximately, the point at which urbanization and mechanization virtually killed what remained of rural American spirit and vitality. Not more than 120 years before, a visitor to America would have found a decentralized country with relatively self-sustaining regions, each with its distinct culture. A village on the farming frontier that wanted to survive had to have enough raw materials to avoid having to import the majority of necessities, and it had to have the industries to transform these raw materials into usable goods. At this stage of the frontier, coming after the Indian trader and trapper, a minority of items, such as steel plows and firearms, were imported. The blacksmith, however, still fashioned harnesses, the wheelwright still crafted wagon wheels, and farm women still made dresses out of flour sacks, but as population increased and towns grew in size, so did a demand for sophisticated goods.

By 1880 Montgomery Ward's first general merchandising mail order catalog—the world's first—had been circulating for eight years. It and its later competitor, the Sears Roebuck catalog, targeted rural areas. By virtue of the fact that their mass marketing and volume sales made it possible for them to undersell small retailers, both threatened small town economies. In 1946 the Grolier Club, a book collectors' society, chose the Ward catalog as among the 100 books having the greatest influence on American culture. A club prospectus said: "The mail order catalog has been perhaps the greatest single influence in increasing the standards of American middle-class living. It brought the benefit of wholesale prices to city and hamlet, to the crossroads and prairie; [and] it inculcated cash payment against crippling credit . . ." By 1880, national marketing from several sources was introducing a wide variety of products to rural Americans coast to coast and marked the emergence of the national economy. From then on centralization of production increased dramatically, and rural areas, which made up the larger portion of the continent, became dependent on cities, not only for industrial production but for financing.

1880 not only marked the onset of the decline of regional economies, but the decline of those rural populations in the earliest portions of the Midwest frontier. A comparison of a four-county area in northeastern-most Iowa in the late nineteenth and late twentieth centuries will make the point. The population of three of the four counties peaked in 1880 at 94,816. One hundred ten years later, in 1990, the population had dropped to 75,599. In 1880 the area had 86 flour mills and 28 lumber mills. By 1997 the region did not have a single grist or lumber mill and only one mixing mill. The period from 1880 to 1900 was the period of the region's greatest economic diversification. Commercial fishing and logging flourished in its river towns, and two of these had clamming industries which supplied shells to local button manufacturers. Sawmills and flour mills dotted the region. Building materials came not only from local timber but from a brickworks and numerous limestone quarries. The fact that virtually all farm families had their own vegetable gardens and butchered their own meat and poultry, meant that they were far more self-sufficient than they are today. Now, slightly over 100 years later, the lumber and grist mills are gone and only a few commercial fishermen hang on. The one remaining button company no longer makes buttons but cards ones imported from Japan. The brickworks has long since disappeared, and the limestone quarries are used solely for obtaining gravel for county roads. As elsewhere in America, supermarket vegetables are imported from California and Central and South America, while meat arrives from out-of-state processing plants. Needless to say, women no longer make their family's clothes, few keep chickens or even grow vegetables and fruits, and fewer still can or freeze what they grow. Long before the year 2000 self-sufficiency had been on the wane, and in the 1980s the confidence that rural peoples had once felt in their societies and future plummeted with the farm economy.

In sum, the early forces of centralization, including large urban financial institutions, mass production, transcontinental transportation, and national marketing and distribution, combined to induce a population shift from rural areas to urban centers. At the same time, the growth of agricultural efficiencies lessened the need for a large force of farmers to cultivate the crops and livestock needed to feed urban masses.

CHAPTER TWO
THE JEFFERSONIAN IDEAL

Today the small-to-medium size farmer knows that his years are numbered, and he knows that farm lands across the country are being transferred into the hands of fewer and fewer owners. In the late eighteenth century, however, matters stood differently. In 1790 farmers comprised 90 percent of the U.S. labor force; by 1850 that number had shrunk to 64 percent. The growth of urban centers and industry demanded ever greater numbers of workers and artisans, resulting in a continuing sharp diminution of farm numbers over the next 130 years. By 1980 farmers comprised a mere 3.4 percent of the labor force; ten years later they were down to 2.6 percent.

Even with a declining number of farmers in the nineteenth century, the farm sector was still able, through mechanization, to produce a surplus of crops. Even with the precipitous drop in farm numbers in the late twentieth century, agricultural machinery and techniques allowed farmers to continue increasing overall production. This production amounted to an explosion, beginning at the end of World War II when the United States began considering efficiency as the standard by which to judge agricultural techniques. After the war tractors replaced plows, mules, and horses. Herbicides and insecticides were soon introduced. All this meant that farming was getting "scientific," along with the rest of efficiently run businesses.

Tractors and chemicals obviously increased costs, but farmers were persuaded to accept them because they decreased labor and increased the chances for greater profits through greater yields. When you think about it, it seems remarkable that organic farming, which farmers had practiced for millennia worldwide, should have been wiped out in a matter of decades. And yet it was, partly no doubt because of the appeal for anything "scientific," and partly because of potential profits. But now one of these elements of "scientific and advanced" farming — the expensive machinery — has become a major contributor to the small farmer's demise.

Even before the effect of machinery's high cost began to take its toll, farmers were persuaded to get big, to buy more land, and to plant "fence row to fence row." In the 1970s FmHA loans were easy to get, so farmers got big. They bought more land, added to their herds, maybe built a new milking barn or added to their farrowing operation. Then in the late 1970s, nobody knows quite why, loans were sometimes called in or rewritten, sometimes underhandedly. A farmer might be asked to sign an agreement that put him out of business.

By the mid-1980s more than a handful of farm families were living through the winters without heat and with very little food. Many watched their herds die of starvation. The strain cracked many. Divorce increased. And then came the suicides.

2.

All this continues to lead our country away from its rural roots, into an ever stranger and more complex future, far from the agrarian vision of Thomas Jefferson, who wanted America filled with farmers because he believed that they are "the most virtuous and independent citizens."

Rather than see Americans divided in employment between manufacture and agriculture, Jefferson wanted to leave manufacturing to the Europeans, for the United States, he thought, could purchase needed goods from Europe in exchange for American food surpluses. One of the biggest arguments in favor of such an arrangement, he argued, was the physical and moral superiority "of the agricultural, over the manufacturing, man."

To John Jay he wrote: "We have now lands enough to employ an infinite number of people in their cultivation. Cultivators of the earth are the most valuable citizens. They are the most vigorous, & they are tied to their country & wedded to it's (sic) liberty & interests by the most lasting bonds. As long as they can find employment in this line I would not convert them into mariners, artisans or anything else."

But on the day Americans become too numerous for the land, then, Jefferson thought, "I should then perhaps wish to turn them to the sea in preference to manufacture, because comparing the characters of the two classes I find the former the most valuable citizens. I consider the class of artificers of a country as the panders of vice & the instruments by which the liberties of a country are generally overturned."

By "artificers" he means the makers of goods, artisans and manufacturers, those who employ themselves alone and those who employ hundreds. What is decisive for Jefferson is that manufacture breeds a demand for luxuries, and is opposed to frugality, a civic virtue. He had in mind the examples of the ancient world, particularly Rome, where a tough and free people acquired a wealth and luxury which corrupted them to the point where they abandoned their liberties for a dictatorship. Even in the early days of the republic, Jefferson considered "the extravagance which has seized them (Americans) as a more baneful evil than toryism was during the war." If Jefferson thought Americans were corrupted then by luxuries, what would he say to us today?

3.

The framers of our constitution understood the intimate connection between economics and politics, between money and political power. "[Alexander] Hamilton and his school," historian Charles Beard wrote, "deliberately sought to attach powerful interests to the Federal Government. Jefferson clung tenaciously to the proposition that freehold agriculture bore a vital relation to the independence of spirit essential to popular rule." Hence Jefferson's passionate desire to see America's lands filled with freehold farmers.

In 1821, not many years after Jefferson's presidency, American statesman Daniel Webster wrote: "It seems to me to be plain that, in the absence of military force, political power naturally and necessarily goes into the hands which hold the property." The early English settlers of New England, he wrote, "were themselves . . . nearly on a general level in respect to property. . . . Their situation demanded a parceling out and division of the lands, and it may be fairly said that this necessary act fixed the future form of their government. The character of their political institutions was determined by the fundamental laws respecting property . . . . The consequence of all these causes has been a great subdivision of the soil and a great equality of condition; the truer basis, most certainly, of popular government."

Webster, then, agreed with Jefferson that popular government rested upon the wide distribution of land among its citizens. Jefferson went further, wanting an agriculturally based economy for the country, because farmers were the best of all classes to uphold their liberty, first by their independence, and second by their lack of corruption. But the powerful and emerging mercantile, banking, and manufacturing interests in the developing country would eventually subvert that dream. Speaking on behalf of those interests, Alexander Hamilton wrote: "The prosperity of commerce is now perceived and acknowledged by all enlightened statesmen to be the most useful as well as the most productive source of national wealth . . ."

Still, the dream of an agrarian society as the underpinning of American democracy had numerous supporters throughout the nineteenth century, including social reformers concerned with the condition of workers in the eastern cities. Cries for distribution of western land began as early as the 1820s, and in 1841 Missouri Senator Thomas Hart Benton, champion of the western farmer, backed the Pre-Emption Act, which allowed farmers to claim unsurveyed land, work it, and eventually buy it from the government. By the mid-1840s, pressure to open western lands intensified when Ireland's

Great Potato Famine caused millions of Irish to seek deliverance from starvation by emigrating to America. These and other immigrants began crowding eastern cities, swelling the labor force, exacerbating an already critical situation. An obvious solution to the misery of those workers was to open up western lands for cultivation, but the move was opposed by northern industrialists, who wanted to retain a hold on cheap labor, and by southern planters, who feared that the western farmer might oppose the creation of more slave states. Nevertheless, it was a mid-South politician, Andrew Johnson, who in the 1840s proposed free distribution of public lands. The proposal was effectively blocked, but a decade later in 1854 the Kansas-Nebraska Act put more western land up for sale.

Johnson's proposal, however, was far from forgotten, and in 1860 the Homestead Act was enacted by Congress but vetoed by President Buchanan, who thought it smacked of communism. Nevertheless, in 1862 the Homestead Act was signed into law by Abraham Lincoln. The act allowed that ". . . any person who is the head of a family, or who has arrived at the age of twenty-one years, and is a citizen of the United States, or who shall have filed his declaration of intention to become such, as required by the naturalization laws of the United States, and who has never borne arms against the United States Government or given aid and comfort to its enemies, shall, from and after the first of January, eighteen hundred and sixty-three, be entitled to one quarter-section or a less quantity of unappropriated public lands . . ." For a ten dollar filing fee a man or woman meeting the above requirements could purchase 160 acres. He or she had to live on it for five years, build a dwelling, dig a well, till the soil, and fence a portion of it. At the end of the five years (and in some case less than five) that quarter section would be theirs.

In exchange for this intended relief to industrial workers, the Federal Government instituted the Immigration Act of 1864 to benefit industrialists. This act, wrote Charles Beard, "gave federal authorization to the importation of working people under terms of contract analogous to the indentured servitude of colonial times." With it, industrialists would receive the labor they needed to replace the workers who left the cities for quarter sections of land.

But the homesteader faced several obstacles: in the first place, he did not need any farming experience to qualify, and many who made the attempt lacked the knowledge to win clear title; in the second place, the quarter section of land was, in arid regions, too small to sustain a farm. Many of the claimants packed up and left, while others, who had acted on behalf of speculators, handed over their lands in exchange for cash. Even before the Homestead Act, much of the Midwest's best agricultural land had been deeded by the federal government to the railroads. Still, despite the failure of many homesteaders, between 1867 and 1900 about 600,000 people had received clear title to homestead lands totaling 80 million acres. Between 1900 and 1915, another 344,444 won clear title.

Although large numbers benefited from the Homestead Act, there never was sufficient population shift to force the raise of urban wages. Historians Fred Shannon and Ray Allen Billington, a student of Frederick Jackson Turner, both thought it failed in this respect. Billington wrote that "the laborers the act was supposed to help didn't have the money to move to the frontier and buy farm equipment." And according to Shannon: "George Henry Evans and his fellow agrarians . . . harped incessantly on the issues of widespread misery, poverty, and unemployment as a consequence of capitalism and land monopolies. All this they confidently expected to be remedied by a homestead policy which would give land to all who could use it."

If it failed to result in equity for urban workers, the Homestead Act laid the basis for a prosperous agricultural society in the Midwest that lasted over a century. But that society was unaware of the consequences of centralization, and as the decentralized regional economies were progressively absorbed into a national economy, the country's monied interests gradually extended their reach into rural America. By the 1920s commercial and financial interests had triumphed to the point where Calvin Coolidge could proclaim that "the business of America is business."

4.

When we examine traditional civilizations, we find that one thinker after another warns us of the perils of commerce and finance. The argument, though, is from a different point of view than that of Jefferson, who is himself echoing the fears of Roman stoics. The stoics saw that wealth and luxury had corrupted the Roman people, but for Plato and others the argument against business revolved around its ability to corrupt the arts. For the Greeks and other traditional peoples, art was not confined to painting, sculpture, music, literature, and dance. The word 'art' itself is our clue to that, deriving as it does from the Latin word 'ars,' meaning skill, trade, or profession, as well as 'art' in our restricted sense. Thus in traditional societies anyone who made a thing was an artist. So were those who nurtured, such as physicians and farmers.

Underlying the very foundations of traditional societies was the knowledge that to lead a fully human existence a person must have an art that he follows all his life. The distinction between work and labor lies in the fact that work is imbued with art, shaped by it, and labor lacks art. Strip a person's livelihood of art, and you strip him of his humanity. A person stripped of his humanity eventually turns to violence, and much of the anger in this country comes from people working jobs that are better suited to robots than to human beings. As for farmers forced off the land, most are obliged to trade a complex art with multiple activities and skills, for low-skilled labor.

5.

To understand how fully the deck is stacked against farmers you must understand that with few exceptions the only farmers who stand a chance of crawling out from under debt are those who can somehow market their own products directly to consumers or sell through farmer-owned cooperatives; otherwise they are locked into a price determined by the four or five vertically integrated agribusiness clusters, each of which is allied to one of the four major international grain corporations, two of which are U.S. based. These multi-national corporations are the purveyors of wheat, corn, soybeans, rice, and other grains to governments around the world. Indirectly these companies determine, here and abroad, the price of livestock and poultry, as well as bread, cereals, pasta, and other grain based produce. It is these companies, not the U. S. government, which sells U. S. wheat to Russia, Korea, and elsewhere.

In years when the sales of the multi-national grain corporations to foreign governments are relatively low, their influence is lessened, while that of the futures markets in Chicago, Minneapolis, and Kansas City is increased. But the major grain corporations have independent brokers buying and selling futures contracts for them at these markets. Considering the volume at which they buy and sell, their influence is considerable.

The prices on grain, pork, beef, and other commodities can vary widely within a day, affected by weather, scarcity, foreign sales and other factors. The middlemen, speculators, never actually see what they are buying or selling, and most fail to make a profit, but those who succeed can make a fortune. Such middlemen are unnecessary, and are symptomatic of a society whose driving force is avarice. As R. H. Trawney wrote in *Religion and the Rise of Capitalism*, medieval social theory condemned "the speculator or the middleman, who snatches private gain by the exploitation of public necessities."

Those farmers who are willing to play the very involved game of puts and calls on the commodities markets may protect themselves from loss. But the majority of farmers, nine out of ten, do not speculate on the market, and do not want to. It is not in their nature. Yet the politicians, the bureaucrats in the U. S. Department of Agriculture, and the bankers are expecting the farmer to become a "good manager," which means they expect him not only to use a computer to record his yields, profits, and losses, but to operate successfully in the futures markets. But the farmer is a special kind of artist, and to ask him to abandon his art and take up someone else's is to expect him to violate his nature.

This brings us to the very heart of this "civilization's" malaise: the denigration and abandonment of vocation, which is intimately connected to the idea of art. In traditional civilizations, all people had a vocation. A vocation is a calling to this or that kind of work, and this calling is determined by our aptitude, which directs our love. This is to say that we love what we do well or what we are called to do. But few remember the idea of vocation, or if they remember it, dismiss it, for we are a pragmatic people, and as pragmatists we can see no difference between work and labor.

Today the values of the commercial class are those that drive this society and its institutions. One outcome is that today, as in Plato's time, commerce has infected all the arts. Artists of various kinds, physicians, surgeons, and lawyers among them, confuse the art of making money with their own special arts. The small farmer has resisted, asking only for a fair price and the opportunity to continue farming.

But the pragmatist has no use for the small farmer, who is inefficient. He is inefficient because he is not a "good manager." And being inefficient, he is undesirable. As former Secretary of Agriculture Earl Butz has said repeatedly, "There are too many farmers." But what this society has yet to learn is that efficiency is a totally inappropriate standard by which to judge human beings and their work, though an appropriate one for robots.

To work backward, the farm crisis can be seen as the final clash between the urban forces of commerce and banking on the one hand, and agrarian, democratic interests on the other. The issue of the contest is not much in doubt, and when the farmer and his way of life pass on, the fiber of American democracy passes with him.

6.

There is a beautiful photo book, *Neighbors*, which is a forty year record of farm families in Jo Daviess County, Illinois. In it, one of the farmers says. " . . . I love this land, all right. To me the land is my being. It's all I've got. It's my existence. I feel like I'm just a part of it. When you read in the Bible where it says God gave you this land to till it, to take care of it, to prosper, that's what it means to me. It's my duty to do this. I don't consider it a job exactly. It's a duty. A responsibility. That gives me happiness and satisfaction and a reason for being here."

How many of us can say that we have a responsibility to do our work, beyond the responsibility to provide food and shelter for our families? For the vast majority of us, our work means nothing more than a paycheck. We have not found what the Buddhists call "right livelihood." We have not found our vocation, so do not know what it is we are supposed to do, have not found responsibility, and consequently remain irresponsible. But a society composed of people without duty and responsibility has no human-centered course or direction. Without duty we are alienated, and that accounts, in part, for why so many of us are angry, why there is so much violence.

CHAPTER THREE
THE RISE OF THE MERCHANT CLASS

We cannot understand the nature of contemporary agriculture until we recognize that it is a consequence of the West's rejection of tradition. That is, the very concept of industrialized agriculture could not have arisen, let alone have come to dominate world agriculture, before the principles underlying a traditional civilization had been discredited. If we seek to renovate agriculture, we must recognize the principles underlying our arts and institutions, for renovation depends not just on remedying particular ills, but repudiating the cause of illness.

When examining our culture as a whole, the first thing to note is that the master idea governing it is economics. From the perspective of economics, all arts are expected to subserve the art of making money. The transformation of economics from an art supporting human life into the master art governing the entire culture, indicates the transformation of avarice from a vice to a virtue. Avarice, of course, did not grow by happenstance: avarice and the other elements of humanity's shadow side could not develop into such proportions until ideals, which once regulated human behavior, had been discredited.

Until the fourteenth century, the dominant philosophy in the West was realism, by whose account reality lies in the immaterial realm. In the West this reality was expressed in Christian cosmology and Platonism, whose essence lies in the claim that all species, plant and animal, as well as minerals and even human artifacts, have immaterial prototypes or forms.

Once realism was discredited, the path was cleared for empiricism, which claims that all knowledge derives from sensory experience. Empiricism is the philosophical expression of the merchant class mentality, and it is no coincidence that at the time of realism's defeat, the merchant class was a powerful force in Western civilization. The merchants, of course, were not responsible for the defeat of realism, but realism's displacement and the merchant's growing power are evidence of a shift in European focus. Prior to the twelfth century, the Catholic Church had kept merchants in check by teachings on just price and just wage, and by proscriptions against usury, or the unjust levying of interest on borrowed money. The man who made his living not in making or praying or administering, but in selling, was a man whose soul lay in peril of damnation.

The Church's teachings on wage, price, and usury held greed in check so long as Europe remained in a barter economy. From the ninth through the twelfth centuries, there was little need of coin. Most Europeans of those times lived on manors, which were largely self-contained, closed agricultural systems. Indeed, at that time medieval Europeans did not have an idea of economics as we understand it. Exchanges were made not only to provide necessities, but also used to reinforce existing bonds between people.

Now the shift from a barter or natural economy to a monetary economy also marked a shift from an agrarian society to a commercial one, and in the process the Church itself became corrupted. As the here and now came to have increasing significance and weight, scientific investigations multiplied, and by the seventeenth century the West saw the appearance of scientists and mathematicians of the caliber of Newton, Leibniz, Galileo, and Huygens. By the eighteenth century, science (and through it technology) was heralding ever greater promise of material prosperity. Merchants and merchant values were on the verge of dominating all spheres of activity.

It was in the eighteenth century that the idea of progress was born. The heart of the doctrine of progress lies in the belief that both human knowledge and material prosperity will continue to expand indefinitely into the future. Under the idea of progress, the end of life is implicitly assumed to be the accumulation of goods, money, and temporal power—all worldly preoccupations. In the twentieth century the material world received such focus that economics came to dominate all other social and political ideas, and today we explain every activity, ultimately, in terms of how much it earns or how it serves the economy. For the last five decades the merchants in charge of this culture, whether businessmen or bankers, or even presidents of land grant universities, have spared no effort to get the farmer to think in terms of profit and loss, and to regard himself as a manager; in short, to think of agriculture as a business. To effect this change merchants have tried to get the farmer to abandon a natural sensibility for a mechanical and urban understanding, to think of nature as rationalists do, as composed of discrete building blocks that can be moved and rearranged at random.

A television ad for a chemical poison, for example, showed a virile young man in sweat clothes jogging between crop rows in the midst of a great field, saying, "I like to examine my factory without walls." The young man was clearly an urban clothing model, but the manufacturer believed that farmers would want to identify with this sophisticated, well groomed and virile young man. To underline the modernity of the concept, the ad was shot in black and white with camera angles that conveyed a sensibility that is up-to-date. The idea is that farming, to be efficient, must be up-to-date, and being up-to-date means thinking of farms as factories.

The idea of progress accustomed us to innovation, specifically technological innovation, and as time went on, to new product development. We came to expect it, in all fields. It became an essential component of capitalism. And now, in order to maintain profits, all industries, including agriculture, have to innovate constantly. It is now a truism that the first companies to adopt a new technology or product that succeeds will be the ones to dominate that market and make an enormous profit. The others will be losers. Thus capitalism itself became caught on a treadmill, and the public had to be conditioned to consume without respite. Expensive farm machinery and the latest pesticides became a necessity for those who wanted to maximize efficiency and remain competitive with other farmers. In order to keep the farmer on the treadmill it became necessary for business to convince the farmer that he must be progressive, i.e., scientific and businesslike, and must compete with other farmers.

2.

All of the governing virtues of capitalism are, as I have indicated, opposed to those governing traditional societies. Most traditional societies are marked, in part, by the fact that they are hierarchical. Unfortunately, there is a strong bias among "progressive" thinkers against hierarchy, a bias which is not only misconceived but actually aids the perpetuation of the present power complex by introducing confusion and in some cases resentment against social organization itself. But the present social inequities are not the inevitable by-product of hierarchy, but of a hierarchy in which people of limited apprehension have taken control of entire societies. For this reason it is necessary to look at traditional societies in which the social organization reflected the capacities, and therefore the responsibilities and duties of the various classes.

Traditional societies are of two types. The first is either a hunting and gathering or primitive agricultural society, and has few specialized roles. It may be democratic or hierarchical. The second type is also pre-industrial, but clearly hierarchical and far more complex economically and culturally. The first type is represented by archaic peoples such as nineteenth-century Native Americans, the various African tribes, the Aurunta of Australia, and so on. The second type is represented by such cultures as that of Vedic India, Confucian China, and medieval Europe between the tenth and thirteenth centuries.

There is a body of opinion now which holds that archaic societies alone are traditional; but this springs from a desire to remake society upon archaic models and thus, presumably, make them more democratic and politically correct. These critics fail to distinguish between the theoretical basis for a social structure and its inevitable corruptions. It presumes that one type of social organization is best for all populations regardless of their size, the availability of natural resources, their region's geography, topography, and so on. It ignores the very elements that have been infused into the fabric of its culture and thus help differentiate it from others.

Writer Kirkpatrick Sale is one who claims that archaic societies alone are traditional. They are, he claims, without "organized stratifications," or what he calls "structures of domination." He further claims that while these societies have different roles and specializations, and while "the individuals may gain status and admiration for particular successes, they do not occupy higher or lower positions among their colleagues."

His arguments for the democratic nature of traditional societies are based on a generalization of Native American cultures, a generalization which fails to describe the social organizations of all or even the majority of other preliterate cultures, including Native American societies themselves. African tribes, for example, have kings and Native American tribes have medicine men, both of which function in war and peace. And by the fact that they fulfill necessary and on-going functions for their tribes, their positions are institutionalized. Clearly, kings and medicine men not only have higher status than others in their respective tribes, they also have positions of authority; i.e., they possess power. Furthermore, to claim that preliterate societies are without structures of domination is likewise incorrect. In some preliterate societies, for example, females are disposed of at the whim of the males, as when the men of the Australian Arunta deflower a friend's bride-to-be and then have intercourse with her, or the Yanomano of the Amazon display their manhood by beating their wives to warn off other males.

As we live in the realm of manifestation, not in that of

the archetypes, corruption inheres to some degree in any society. One advantage of the traditional hierarchical society over contemporary Western societies is that traditional hierarchies are based on a theory of mutual duties and responsibilities, whereas in the West the culture of individualism does not recognize an individual's obligations to society.

This is to say that traditional hierarchy is based on natural castes (those based on aptitudes and capabilities), not social castes. Of course, social castes have at times supplanted natural castes and natural castes have periodically degenerated into social castes. But the hierarchy of natural castes is based on the knowledge that each of us has gifts, some more crucial than others to the state or tribe's well being. Certain work obviously can be done by anyone of sound body, whereas wise political deliberation is neither innate in all nor an art that everyone can acquire. A little reflection makes clear that gifts of the mind and soul have primacy over gifts of the body, because action without forethought can be disastrous.

In explaining social hierarchy, traditional people made use of the metaphor of the body, likening its parts to the different castes and classes of society. The sacred Hindu book, *The Rig Veda*, records the origin of the four major castes. They sprang, it is said, from the body of the god Purusha, who was offered up for sacrifice by the other gods. The scripture says: "His mouth became the brahman; his two arms were made into the rajanya; his two thighs the vaishya; from his two feet the shudra was born." The brahman or priestly caste, whose words are based on knowledge, speaks on behalf of the body. The rajanya or nobility create and maintain the social order envisioned by the brahmans. They are the warriors and administrators of the realm. The vaishya—the merchants and farmers— are the thighs, a necessary support, but higher in the hierarchy than the shudra, or laborers, who are the feet. From *The Laws of Manu*, another ancient Hindu text, we learn that "for the sake of the preservation of this entire creation . . . [Purusha] assigned separate duties to the classes which had sprung from his mouth, arms, thighs, and feet." Because these castes are formed from the body of Purusha, the order of castes is sacred and in the best interests of all. The castes, as the metaphor of the body implies, are mutually interdependent.

A medieval text, the *Policraticus* by John of Salisbury, states the medieval position almost identically. "A commonwealth . . . is a certain body which is endowed with life by the benefit of divine favour, which acts at the prompting of the highest equity, and is ruled by the moderating power of reason." The clergy represent "the soul of the body" and the prince represents its head, and is subject only to the clergy. In John's time, European merchants were so few in number as to be negligible, and farmers—the peasantry—were in a class by themselves, which he likened to the feet. The order of classes is the same for Hindus and medieval Europeans: the priests have authority over the nobility, who have authority over the rest of society. While the Hindus had four castes, medieval society was composed of three estates. Since all classes or castes are necessary, traditional societies emphasize the duties and responsibilities of each. Without these, the body politic devolves into chaos.

We find the idea of vocation and gifts at the heart of Christian teaching. In *I Corinthians,* 12, Paul likens the Church—the sum of believers—to the body of Christ. Just as "the eye cannot say unto the hand, I have no need of thee: nor again the head to the feet, I have no need of you," just so do the members of Christ's body need one another. "And God hath set some in the church, first apostles, secondarily prophets, thirdly teachers, after that miracles, then gifts of healings, helps, governments, diversities of tongues." For Paul, as for the Hindus and medieval Europeans, there exists a hierarchy of gifts and vocations.

What then separates us into different vocational classes? Love for a particular form of work is one. But that love reflects the individual's apprehension of what is Real. The four Hindu castes can be divided into two groups. For the brahmans and the rajanyas what is Real is the noumenal world, the non-material world of Mind and Ideas. For the vaishyas and shudras, Reality resides in the phenomenal world, the world that is sensed. For brahmans contemplation is the highest activity, and through contemplation they apprehend the nature of things. For the rajanyas the highest activity resides in instantiating and maintaining the order apprehended by the brahmans. For the vaishyas the highest activity is collecting goods or working with one's hands. For the shudras the best activity is that which produces the greatest pleasure: sexual activity, eating, and drinking. In our time the classes have become mixed and values inverted. The brahmans and their values are ridiculed, while the vaishyas—the merchants—are the leaders. Pragmatism and philosophical materialism are the philosophies of the merchant class, and it is these philosophies which have determined the activities of this society for several centuries, at least. In the course of its triumphal procession, business has drawn to its ranks not only those whose mentality is perfectly atuned to it, but many who have simply lacked the courage to follow the promptings of their hearts, as well as those who feel no call at all. But among the ranks of businessmen and entrepreneurs, it must be noted, are a few who have not only used their wealth for the betterment of their fellow beings, but created it for that purpose.

The dominance of merchant values has entailed a consequent weakening and destruction of those virtues by which society is maintained. The absence of self-restraint in a culture that encourages the individual to fulfill any and all appetites has been fertile ground for the growing power of the shudras, whose criminal organizations in the United States, Russia, and Japan extend into major legitimate businesses.

3.

Traditional societies did not hold the merchant in great respect. His aim was profit, but he made nothing. While others worked, the merchant profited by their labor. Plato had acknowledged the need for merchants and shopkeepers when he set about devising his ideal state, but thought that "in well-conducted cities they are generally those who are weakest in body and who are useless for any other task. . . ." In *The Laws* he wrote that a maritime city is filled as it is "with unwholesome traffic and retail huckstering [which] breeds shifty and distrustful habits of soul, and so makes a society distrustful and unfriendly within itself as well as toward mankind at large."

In the early days of its republic, Rome was a society of frugal and stoic farmers, and therefore had little use for merchants and their luxuries. Only after the republic was supplanted by empire did Romans indulge in ostentatious display. Plato had observed the same corruption penetrate Athens as it became an empire. And the Catholic Church, seeking to focus people's minds on the state of their souls, warned that men engage in commerce at the risk of losing paradise.

4.

The rise of the merchant to the dominant position in our civilization was 700 years in the making. In Europe, in the early Christian era, the merchant was virtually non-existent. In the second century, Europe's population began to decline, and in the ninth reached its nadir. Famine and disease did most of the work. Tribal invasions, which began in the third century, continued sporadically into the twelfth. Peasants put themselves under the protection of war lords, and monasteries proliferated as people banded together for safety and a life of prayer. The cloister became the one stable structure in this time of disorder and confusion.

By the tenth century, Europe's population began to grow. Raids were less frequent, and the raiders were settling throughout England and Europe, and converting to Christianity. Starting in the eleventh century, towns were created by merchants clustering about castles or monasteries, under the protection of a lord or abbot. Forests were cleared, more land was tilled, more crops were grown. As agricultural production increased, the population swelled, no doubt due in part to a feeling of increased security. Towns became centers for local trade: for the barter of grains, livestock, wine, and basic tools. Luxury items appeared later with the growth of international trade, carried out at specified markets towns, at specific periods.

5.

For centuries the Church had taught that one did business at the risk of one's soul. Jesus had said: "No man can serve two masters; for either he will hate the one and love the other, or else he will stand by the one and despise the other. You cannot serve God and Mammon."

By the twelfth century, during the rise of trade, scholastics were writing systematic expositions on trade and usury. English historian R. H. Tawney wrote: "The movement which prompted them"—the expositions—"was the growth of trade, of town life, and of a commercial economy, in a world whose social categories were still those of the self-sufficing village and the feudal hierarchy."

The village and the hierarchy were threatened by the growth of trade, by avarice in particular. Avarice corrupts the soul and destroys the individual, but its effects go beyond the individual to the body politic. By the eleventh century coins were being minted, and an emerging monetary economy was displacing the barter economy. This new economy threatened Christian values and social stability. Avarice, as the chief threat to stability, became the chief of the Seven Deadly Sins.

People of the Middle Ages lacked our conception of economy. As Karl Polanyi wrote: "We must rid ourselves of the ingrained notion that the economy is a field of experience of which human beings have necessarily always been conscious. To use a metaphor, the facts of the economy were originally embedded in situations that were not in themselves of an economic nature, neither the ends nor the means being primarily material." Exchanges existed to strengthen social relationships, and the notion of reciprocity dominated medieval man's theory of economic exchanges. People drew relationships closer through an exchange of gifts. One party gave something to a second party, and the second party gave it back, albeit in a roundabout way. The purpose of the exchange, wrote historian Jacques Le Goff, was "to draw relationships closer by strengthening the ties of reciprocity."

Starting from the practice of reciprocal exchanges, it is not difficult to see why medieval people despised the merchant. Medievals drew a line between the craftsman who labors for his living, who makes something, and the merchant who merely sells a thing as he buys it. The medieval monk Gratian had written: "Whosoever buys a thing, not that he may sell it whole and unchanged, but that it may be a material for fashioning something, he is no merchant. But the man who buys it in order that he may gain by selling it again unchanged and as he bought it, that man is of the buyers and sellers who are cast forth from God's temple."

From the traditional perspective, amassing a fortune does not earn a person respect. And creating immense holdings for oneself while destroying a community is abhorrent, at the very opposite of what exchanges are supposed to do.

6.

Ever since the Fall, man was meant to labor as penance for his sin. Labor was honorable and could lead to redemption. As we read in *Genesis*: ". . . in the sweat of thy face shalt thou eat thy bread." Since the merchant, according to early medievals, does not labor, either mentally or physically, he exempts himself from God's ordinance. The usurer was placed among the merchants: he was, in fact, the lowest of them. His work is iniquitous, and he gathers wealth by unjust means. Usury is a form of interest, the product of a transaction that should not have produced profit, for nothing was made or transformed. Jacques de Vitry, a medieval lay preacher, had written: "God created three types of men: peasants and other laborers to assure the subsistence of the others, knights to defend them, and clerics to govern them. But the Devil created a fourth group, the usurers."

Beginning in the fourth century the Church had forbade usury, first to clerics, later to laymen. Between the fourth and the twelfth century, with few coins circulating in Europe, usury was not widespread. But in the thirteenth century, the monetary economy emerged as trade began to flourish; so did money lending. The Cahorsins, Christians whose name derived from their home city of Cahors, practiced their trade throughout Western Europe and made their name synonymous with usury. At the same time the Lombards were also lending money across Europe and became noted for the excessive interest they charged—to the extent that one grant to moneylenders forbid them to make the demands that were customary with the Lombards, whose usual interest was 43 1/3 percent. Medievalist Henri Pirenne wrote: "Compared with the efflorescence and ubiquity of Italian credit, that of the Jews appears a very small affair and the part they played in the Middle Ages has certainly been much exaggerated. In actual fact, the more economically advanced a country was, the fewer Jewish money-lenders were to be found there."

With the appearance of the monetary economy, borrowing and lending increased, as nobility and merchants alike sought credit. Up to that time, technology and agriculture had been steadily improving, while townsmen and peasants had become increasingly prosperous and more worldly. As usury grew, the Church sought to stem its growth: the Third Lateran Council in 1179 declared that the social order was threatened by the number of people leaving their trades to practice usury. Two centuries later so many peasants were leaving the land to take up usury that the Church feared that famine was imminent. Nevertheless, by the fifteenth century the Church itself was becoming addicted to pomp and ostentatious display, to costly art and architectural projects. In such an environment, the usurer found himself increasingly needed and accepted.

In time, scholastics devised five excuses for usury in commercial activity. These reasons did not, however, extend to consumer loans, for which the usurer would still earn damnation. But the idea of Purgatory, developed sometime around 1000, offered him hope, as did the changing concepts of sin and penance. Between the late eleventh and the early thirteenth centuries, sin and penance became spiritualized and internalized. Intention became the means by which to measure a sin's gravity. And penance after the thirteenth century became a private matter between confessor and sinner, whereas before it had been public and collective. Now penance was determined by the sincerity of the contrition, as best the confessor could determine it. Thus, if the usurer intended to make restitution for his ill gotten gains and if he were contrite for his usury, he might find himself in Purgatory. And the soul in Purgatory would eventually gain Heaven. Thus the usurer could have his gold and heaven too.

So we see medieval Europe undergoing the same transformation that Rome and Athens had undergone. Prosperity increased a desire for luxury goods, for more elaborate palaces and churches, and therefore a need for credit. Merchants and usurers flourished.

7.

Money is not simply a neutral tool, a medium of exchange, for with the appearance of metallic currency, a man and his work become commodified. In a barter economy, and even in a proto-monetary economy where an item such cotton clothe (Africa) or rice (Japan) or oxen (ancient Greece) are measures of exchange, the work and workman retain their dignity. Traditional social orders are not disturbed by these measures of value. But the coin marks a disturbance: it introduces instability. This it does first by the fact that work (vocation) becomes measured by something that hasn't the practical value of a bolt of cotton or an ox. Money "objectifies" work: measures it against something abstract. And if a man's vocation or work is objectified or commodified, so is the man himself. Second, because the value of gold and silver fluctuates depending on their relative abundance or scarsity, instability—inflation and deflation—are introduced into the economy. This fluctuation is complicated by the varying value of the coins of different countries or regions, owing to their differing amounts of gold and silver. Inflation drives up the price of necessities and increases the rift between the haves and have-nots.

In the Middle Ages between the ninth and twelfth centuries the circulation of metallic currency was extremely limited, mostly for trade at weekly town markets, and there restricted to small retail purchases. Market trade was for the most part confined to burghers, or town residents. Coins also doubtless changed hands at the great annual fairs, beginning in the twelfth century, in a dozen or so cities where the great merchants met to buy wholesale from one another. (Flemish merchants at the great Troyes fair, for example, might trade Flemish clothe for gold and silver ornaments and spices imported by Italian merchants and sell these to merchant traders on the North Sea coast.) Obviously money in the form of coins is not adequate to handle large transactions: one cannot carry or transport enough gold or silver to stock a warehouse with imported luxuries or to finance a large building project. Even for lesser exchanges, gold and silver are not often feasible. At the fairs, then, credit in the form of bills of exchange played a greater part than the exchange of coins. In time, through necessity, the medieval prince and merchant rediscovered several ancient instruments of credit, including not only bills of exchange, but promissory notes, checks and bank notes, which became the fundamental tools for expanding the economy.

Fernand Braudel points out that large areas of Europe, not to mention other parts of the world, remained untouched by the monetary economy and retained the barter system until the eighteenth century. But where money made inroads, it transformed everyone's world, not only in the price fluctuations of necessities and in the transformation of values, but in the destruction of customs. Money, he writes, is a technique that creates injustice, for "... any society based on an ancient structure which opens its doors to money sooner or later loses its acquired equilibria and liberates forces that can never afterwards be adequately controlled."

8.

The fourteenth century was a time of intellectual and spiritual brilliance. Besides being an era of great logicians and grammarians, it was the time of mystics like Meister Eckhart who taught contemplative prayer and in so doing demonstrated that it opened up intellectual intuition, what he called the passive intellect, that which achieves understanding by its openness, through direct cognition, without the aid of mediating symbols. Intellectual intuition is distinguished from the active intellect, sometimes called the discursive reason, that which speaks, compares, analyzes, and describes, writes poems and scientific treatises, understands by means of symbols, mathematical or linguistic. Intellectual intuition penetrates the noumenal. It enters the realm of archetypes and understands that without the archetypes or ideas there can be no support for the material world. The medievals referred to the theory of eternal archetypes, ideas, or universals as realism.

But sometime in the fourteenth century realism was challenged by nominalism, a theory which asserts that ideas derive from individuals and are names for classes of objects perceived by the senses. Nominalism, wrote Richard Weaver in *Ideas Have Consequences*, brings into question "whether there is a source of truth higher than, and independent of, man; and the answer to the question is decisive for one's view of the nature and destiny of mankind. The practical result of nominalist philosophy is to banish the reality which is perceived by the intellect and to posit as reality that which is perceived by the senses. With this change in the affirmation of what is real, the whole orientation of culture takes a turn, and we're on the road to modern empiricism."

The hierarchy of natural castes is based upon the distinction between what each thinks is real. By the time the nominalist position became generally accepted, the merchant and others for whom experience is the surest guide to truth, were coming into ascendance. The subsequent history of the West can be read as the continued rise of the merchant class and the growing dominance of its ideas in all domains. Its influence extends even to those scientists for whom reality resides in the material world, and for whom the scientific project means the quantification of all phenomena. French metaphysician Rene Guenon aptly named our age the Reign of Quantity for its insistence that quantitative measurement in one form or another describes phenomena in their fullness, while also claiming that their qualitative aspect belongs to the subjective realm and hence is dubious and uncertain.

It is perhaps ironic that as realism was being discredited with Western Europe's shift of focus from the transcendent to the material, that the economy was moving in the opposite direction, from the tangible to the abstract, from barter in goods towards money and credit and their commodification—abstraction—of the individual and his work.

CHAPTER FOUR
FARMING AS A "SCIENTIFIC BUSINESS"

The history of nineteenth-century agriculture shows that isolation and competition have not always been the hallmark of the American farmer. Indeed, American political history of the late nineteenth century demonstrates that cooperation was once a driving force in rural America. That cooperation, however, had to be forged from the arduous and difficult position of the farmer who, after the Civil War, was forced to pay the railroads' exorbitant shipping rates or see his produce and livestock go unpurchased. Furthermore, he had no choice but to pay the prices that suppliers charged, and to accept whatever money buyers offered.

The only remedy for the independent farmer lay in collective organization. In 1867 the Grange emerged, the first and best known nineteenth-century national farm organization. Formed in 1867 by Minnesota farmer Oliver Kelley, the Grange at first had modest economic goals and focused primarily on social and educational activities for its members. By the 1870s, however, it began putting most of its energy into opposing railroads and in forming farmer owned cooperatives. Other state farm groups formed in the late 1870s and eventually amalgamated into three large alliances: the National Farmers' Alliance of the Northwest, formed in 1877, the Colored Farmers' National Alliance and Cooperative Union, organized 1886, and the National Farmers' Alliance and Industrial Union, formed in 1888. These from the outset focused on cooperative economic action.

But other factors besides the power of the railroads

and middlemen combined to make the farmers' position intolerable. From the end of the Civil War onward, the farmer had been drawn slowly but inexorably into the vortex of capitalism. Hard manual work was gradually replaced by mechanized planting and harvesting. In the 1870s and 1880s there came a host of farm machines, including the cultivator, corn picker, reaper, hayrake, corn sheller, seed drill, disc harrow, potato digger, spring-toothed harrow, combine, riding plow, and hay loader. By adopting these and other technologies, the farmer became a capitalist and almost as dependent upon machinery as the manufacturer. And like the manufacturer, the farmer voluntarily, albeit unknowingly, accepted one of capitalism's mandates: the continual purchase of efficiency improving devices, and consequently found himself burdened by ever-increasing debt. For whereas a hand-made implement might last a lifetime and therefore meant a one-time expense, the new machinery needed frequent repair and eventually wore out. But, as the price of grain was determined by that produced on the most productive land with the latest equipment, the farmer was forced to make machinery purchases.

The farmer's absorption into the capitalist system was abetted by the movement away from self-sufficient farms with multiple crops and animals to one or two crops and specialization in either beef cattle, hogs, or dairy cows. The high cost of machinery designed for one crop necessitated the shift to a monoculture: in the South cotton remained king; in the Midwest, corn and wheat dominated. Machinery increased production and by the late nineteenth century Midwestern farmers were not only feeding America but supplying a flood of grains overseas. Slowly the farm that of necessity had aimed at self-sufficiency and produced food primarily for use, was replaced by one that grew monocultures aimed at earning a profit. With that transformation of aim came a whole new psychology and type of individual, although its full flowering did not appear until our own time.

Once he sold for profit and was tied to monocultures, the farmer's fortunes were linked to economic cycles with their periodic depressions and the hazards of foreign trade. Thus the depressions of 1873 and 1893, combined with high tariffs, agricultural overproduction, and a stagnant monetary supply, aggravated the farmers' position, sending more and more into tenancy. In 1880 a quarter of all farmers were tenants; by the early 1900s the ratio had risen to a third. Meanwhile the number of farm mortgages increased, and the younger generation of farm children sought more remunerative work in cities.

Since colonial times, agrarian and commercial interests had battled over currency issues. The 1786 insurrection known as Shay's Rebellion was confined to Massachusetts but was sparked by issues plaguing other states as well. There was, across the confederacy, a shortage of currency, and farmers everywhere were burdened by heavy debt. But Massachusetts was also levying a heavy tax, which went primarily to repay wealthy Boston creditors. This was the added spark that ignited Shay's Rebellion. The insurrection was quickly suppressed and brought increased demands for a strong central government, but the issues for which Shay and his compatriots fought remained.

The currency debate inflamed commercial and agrarian factions up until the Civil War, focusing in part over whether to establish a national bank with a national currency or to maintain state banks and local notes. During the Civil War, the government issued Greenbacks, currency unbacked by specie, gold or silver. Following the war, both factions contended over the Greenback's future—whether or not to make it redeemable, and whether to expand or cancel it. Farmers, burdened as ever by debt, fought against backing it by specie and sought to expand its supply, reasoning that a greater volume of currency would raise prices, allowing them to discharge their debts more easily. But bankers, who held farm mortgages, opposed the move, for inflation would lessen the real value of a dollar and consequently lessen the value of the monthly mortgage payment.

At the same time as they called for an increase in the supply of Greenbacks, agrarian interests were also demanding a resumption of silver coinage (which Congress had abandoned following the 1873 depression) and were demanding it in unlimited supply. During the 1870s, Congress acted on both the Greenback and silver issues, voting on behalf of commercial interests in 1875 to make the Greenback redeemable, and on behalf of agrarian and western mining interests in 1878 by mandating that the government make large monthly purchases of silver to increase the coinage in circulation. The 1878 bill passed because agrarians and silverites joined forces with eastern interests that were demanding high tariffs. Tariffs were another source of trouble for the farmer: high tariffs kept the price of American agricultural implements high by weakening foreign competition.

All these factors, then, combined to make the farmer's plight unendurable. And just as the industrial worker found it necessary to unionize to fight the trusts, so the farmer also realized that his only hope lay in collective action. In 1874, seven years after its founding, the Grange had one and a half million members. It fought for and won legislation regulating freight charges, and it avoided the prices of middlemen by forming cooperatives that sold a wide variety of products to its members. Successful for a time, the cooperatives eventually proved the Grange's undoing. The farmers' Alliances also established cooperatives, but they too failed, principally because of business inexperience. In 1890, the Alliance sought political remedy by forming the Populist Party, some of whose Midwest candidates won state and U.S. Senate races. In 1896, the Populists offered a presidential candidate who won over one million votes, principally in the West, while the Democrats focused on the silver issue and ran William Jennings Bryan against Republican McKinley. In his closing speech at his party's convention, Bryan spoke his now famous words: "You shall not press down upon the brow of labor this crown of thorns. You shall not crucify mankind upon a cross of gold!"

While the silver issue had literal and symbolic meaning for rural Americans, neither New England nor the Midwest responded to it. New England farmers had not experienced the hardships of their counterparts elsewhere in the country, and Midwest cities may have turned their section in McKinley's favor. Even though the farm bloc did not turn out en masse for Bryan, the 1896 election may in fact have marked the crux in the battle between agrarian and commercial interests. Historian George Brown Tindall sees it as the "climactic political struggle . . . between rural and metropolitan America. . . . And metropolitan America had won." Henry Adams, as Tindall points out, had said as much in his autobiography, *The Education of Henry Adams*: "For a hundred years . . . the American people had hesitated, vacillated, swayed forward and back, between two forces, one simply industrial [productive], the other capitalistic, centralizing and mechanical. . . . the majority at last declared itself, once and for all, in favor of the capitalistic system with all its necessary machinery."

No one could avoid noticing the changes that were taking place across rural America, not only in the machinery but in the actions and attitudes of men. Sherwood Anderson had grown up amidst the transformation, and that change lay in the heart of his two prose-poetic works, *Winesburg, Ohio* and *Poor Folk*, volumes which deserve a place on the shelf alongside *Leaves of Grass* and *Huckleberry Finn*. *Winesburg, Ohio* contains a story about a farmer, Jesse Bentley, who, Anderson wrote: ". . . had grown into maturity in America in the years after the Civil War and . . . like all men of his time, had been touched by the deep influences that were at work in the country during those years when modern industrialism was being born. He began to buy machines that would permit him to do the work of the farms while employing fewer men and he sometimes thought that if he were a younger man he would give up farming altogether and start a factory in Winesburg for the making of machinery. . . . The greedy thing in him wanted to make money faster than it could be made by tilling the land."

2.

Despite his reliance upon machinery and chemical inputs (which started in the nineteenth century), the American farmer prospered during the first two decades of the twentieth century. During the 1920s, however, the farm economy took a turn for the worse, and with the Depression it hit rock bottom. The post-World War II boom brought renewed prosperity, and by the early 1950s mechanization was once again proceeding full tilt. It was this mechanization, with its attendant overproduction, that led to the farm crisis of the 1980s.

That crisis was not a phenomenon that just "happened," but was the result of coordinated planning, as far back as the early 1950s, between government bureaucrats, university economists, American Farm Bureau Federation officials, and heads of large corporations. As we examine the development of industrial agriculture, we see a dynamic that began accelerating in the early 1950s and now seems unstoppable. Farmers have been trapped, as it were, in a huge machine that has been slowly, inevitably, grinding up them, their families and surrounding towns.

During the Eisenhower years it was decided that the United States had too many farmers and that millions would have to be forced from agriculture. There were three reasons for this: first, the United States was experiencing a post-war economic boom, which meant that urban industries were expanding and needed more labor. If farmers could be forced from the land, they would undoubtedly migrate to the cities for industrial jobs. Not only would they then fill needed niches, but their increased incomes could be consumed on more American products. Second, at that time diesel driven machinery was beginning to permeate agriculture and was clearly going to map its future. Immediately following the war, affordable tractors became widely available, while engine driven combines began appearing somewhat later, in the early fifties. It was clear that the number of working hours needed to run a family farm were decreasing, and would continue to decrease, even with the increased yields promised by the advent of herbicides and pesticides. For the believers in progress there was absolutely no reason at all to maintain so many unnecessary farmers.

3.

John H. Davis became Assistant Secretary of Agriculture in 1952, and later the Director of the Moffett Program in Business and Agriculture at Harvard Business School. With the advantage of hindsight we note the connection between agriculture and business, a fact that at that time was probably lost on many farmers and the public at large. In a 1956 article that he wrote for *Harvard Business Review*, Davis advocated vertical integration in agriculture—the merger of firms handling different phases of production and distribution of a product or commodity. He called such a movement "agribusiness." "In the dynamic era ahead," he wrote, "the term 'farm problem' will become more and more a misnomer; farm problems will be recognized as being also business problems and vice versa. More precisely, farm problems will be agribusiness problems."

Davis also recommended the elimination of farm programs, including price supports. This was a call that would be heard with increasing frequency in that decade. It was also advocated in the early 1950s by a group of thirteen agricultural economists, including Earl Butz, later Richard Nixon's Secretary of Agriculture. In their report the economists recommended the elimination of price supports, so that "labor resources [that] are being used in agriculture . . . could be used more profitably in other lines." The writers were fully aware that without price supports, many farmers would fail, forcing them to look for other work.

The thirteen economists also recommended that government payments to farmers (payments that maintained parity or equality with non-farm labor), and supplementary payments in case of depression, be eliminated. Their report also recommended that "educational institutions should provide adequate training for workers of all grades and should use their facilities to urge the movement of farm people away from situations where labor returns in farming fall below that in other comparable occupations . . . This principle of voluntary adjustment is basic to our system of free enterprise." The report concluded that ". . . our farm policy needs to be realistically shaped to the conditions developing for the future rather than continue as a legacy from the past." About a decade later all these recommendations would be reiterated by the influential Committee on Economic Development.

From 1942 through 1952 farmers had received, on average, 100 percent of parity. But in 1952 Secretary of Agriculture Ezra Taft Benson tried to get parity repealed. (It was 90 percent in 1952.) Eisenhower replied that "gradualism" was the better policy, but in 1953 the repeal began.

The rhetoric of the period is sometimes distinctly business oriented, at times stressing efficiency, bigness, and profits, while at other times emphasizing an appeal to God and country. Earl Butz, who was another of Benson's Assistant Secretaries of Agriculture, spoke bluntly in business terms, as in 1954 when he declared that agriculture "is now a big business," and that "just like the modern business enterprise, [it] must 'adapt or die.'" Adaptation for Butz meant that the farmer must "get big or get out," a phrase that he used repeatedly over the years. In the same period Butz told the Dayton, Ohio, Rotary Club: "American agriculture's like a big pie. Right now we've got lots of farmers, and each one (is) getting small slice of the pie. We need to eliminate a bunch of them, so that those that are left will get a lot bigger slice." No one could have made the administration's intentions any clearer.

The appeal to God and country was exemplified in an article by Benson, part of which A. J. Krebs quoted in his comprehensive study of agribusiness, *The Corporate Reapers*. "Freedom," Benson wrote, "is a God-given, eternal principle vouchsafed to us under the Constitution. It must be continually guarded as something more precious than life itself. It is doubtful if any man can be politically free who depends on the state for sustenance. A completely planned and subsidized economy weakens initiative, discourages industry, destroys character, and demoralizes the people . . . . The future of agriculture and the preservation of a sound economic system depends on the vigorous reemphasis of the principles, benefits, and values of competitive enterprise. . ."

The farm programs had never demoralized farmers, and Benson had to have known it. Farmers remained an extraordinarily hard working class, even into the farm crisis of the 1980s and after. As for his exhortation to compete, prior to the Cold War no one had thought of farming as a competitive system. Farmers were not trying to drive one another out of agriculture: rural society until fairly recently had always been characterized by cooperation and community. Benson is using Cold War rhetoric to implant business "ethics" in the farmlands. The rhetoric was fairly commonplace. Before leaving his faculty position at Cornell University to serve as Administrator of Commodity Stabilization Service under Benson, James A. McConnell told an audience that price support programs threatened to change the United States into "a state which is predominantly Socialistic in character."

4.

Krebs pointed out that Benson's key administrators and policy makers at the USDA, including Butz and McConnell, and on Eisenhower's National Agricultural Advisory Committee had been on the faculty of Cornell University, originally a land grant college. Today the agriculture schools of land grant colleges, including Texas A & M, Auburn, and Iowa State University, advocate the latest farming technologies and are subsidized by agribusiness giants such as Deere, Pioneer Seed, and Monsanto. In fact, the schools of agriculture at these universities now depend upon agribusiness for a significant portion of their research funding, which calls in question the research itself. Through their county extension agents, the state universities, along with the American Farm Bureau, have been able to promulgate the agribusiness agenda.

The Extension Service grew out of a system of county

demonstration agents, and by 1913 was institutionalized with the passage of federal legislation. Today county extension agents across the country are employees of their respective state universities. One of the primary duties of the Extension Service has been to transmit research conducted by land grant universities and the USDA to the farmer; it has also been charged with the responsibility of helping rural communities with a variety of problems, and with working with rural youth. But a 1970s report on Extension, *Hard Times, Hard Tomatoes,* authored by Jim Hightower, noted: "Like its other partners in the land grant complex, the colleges of agriculture and the agricultural experiment stations, Extension Service has not lived up to its mandate for service to rural people. . . . The focus of ES primarily is on rural 'clients' who need it least, ignoring the obvious needs of the vast majority of rural Americans."

Among other things, the report noted that Extension Service "has pushed high-technology agriculture to the detriment of smaller producers." As Krebs wrote: "Extension agents have acted as salesmen for the agricultural research, research generated by their colleagues in the land-grant college system, which largely serves the needs of corporate agribusiness and larger producers."

But Extension has not only had a long history of collaboration with agribusiness, it has, for years, collaborated with the American Farm Bureau, the self-described Voice of Agriculture and the largest farm lobby in the United States. As its critics point out, the Farm Bureau's 5 million members are predominantly non-farmers, a fact which is readily seen when we recall that USDA statistics show that the United States has fewer than 2 million farmers nationwide, including those earning less than $2,500 in annual farm sales. The Farm Bureau's ranks are filled primarily with those who have joined to take advantage of its inexpensive health insurance; in fact, the majority of Farm Bureau members now live in urban centers and have no physical connection whatsoever with agriculture or rural life.

Despite the income from its $ 6.5 billion insurance business and other interests, the Farm Bureau has managed to maintain its non-profit status, having evolved from the first county Farm Bureau, formed in 1911 in Binghamton, New York, to work with the local extension agent. In 1914 Congress created a national extension service to operate through the state land grant colleges. In order for a county to have extension staff funded by the USDA, it had to have a county farm bureau with a membership of at least 20 percent of the farmers in the county. There lies the origin of the connection between the present day Farm Bureau and the Extension Service.

The Farm Bureau's influence is considerable. According to *Forbes Magazine*, the Farm Bureau has for years ranked among the twenty-five most powerful lobbies in the United States. It has had sufficient clout to quash two congressional investigations into whether its non-profit status is warranted, considering its lucrative insurance operations. The Farm Bureau is powerful enough to have initiated an FBI investigation into two of its critics—a husband and wife who ran a small Chicago bakery and had collected a petition on behalf of the Defenders of Wildlife, which opposed the Farm Bureau's attempts to eliminate wolves from Yellowstone National Park.

Almost since its inception, the Farm Bureau has consistently proposed government policies that favor large farmers, has worked to dismantle policies that work on behalf of the small farmer and the rural poor, has opposed environmental regulations (such as the Clean Water, Clean Air, and Safe Drinking Water Acts) that would better the general welfare, and has consistently advocated the latest agricultural techniques, including the use of chemical poisons.

For example:

—The Farm Bureau worked untiringly to destroy the Farm Security Administration, a New Deal agency whose mandate was to eradicate rural poverty during the Depression. Before the FSA was dismantled in 1937, it came under the usual accusations of being communist, anti-American, and a corroder of self-reliance.

—The Farm Bureau has consistently opposed government support for farmers. The Farm Bureau was an ardent supporter of the 1996 Freedom to Farm bill, a policy that accelerated small farm auctions and foreclosures. Another example of the Farm Bureau's stand in this regard was printed in the April 17, 1999 issue of its newspaper, *The Spokesman*, whose lead article presented a plan by Texas Representative Larry Combest (R), chairman of the House Agriculture Committee, which proposed reducing the then current $130 billion farm income support to a $6 billion fund. While the paper did not officially endorse Combest's position, the article was printed on a green column to highlight its importance.

—The Farm Bureau has consistently opposed environmental protection. The May 22, 1999 issue of *The Spokesman* reported two studies, one by Texas A & M and another by Auburn University, both of which purported to demonstrate that the elimination of two types of pesticides (organophosphates and carbamates) would lower crop yields and therefore increase foreign imports and raise food prices, as well as lower the food's nutritional content.

—The Farm Bureau vigorously supports the World Trade Organization and endorses "Ending the use of all nontariff trade barriers." [*The Spokesman*, 5/29/99] Whether tariffs and quotas are high, low, or nonexistent, the small farmer still receives such low prices that he is often forced out of farming, or forced to take an off-farm job. Those who reap the benefits from the elimination of quotas and tariffs are the agribusiness giants.

—The Farm Bureau has taught insistently that competition is thoroughly American. "Here in America we comPETE!," thundered a Farm Bureau official to Fr. Norm White, Rural Life Director for the Dubuque Archdiocese. Farmers throughout the Midwest and South began reiterating the Farm Bureau line. Thus farmers, who throughout the nineteenth and first half of the twentieth centuries had relied upon cooperation for survival (threshing rings and barn raisings) were brainwashed into believing that cooperation was un-American.

Farmers became isolated from one another, psychologically at least, and when the farm failures of the 1980s began arriving, they were unable to discuss their problems with one another, for each believed that he alone had failed. The Farm Bureau, the banks, and the Extension Service had them convinced that those who failed were "bad managers," that is, poor businessmen. Farming was now strictly a profit and loss operation, and those who could not survive the onslaughts of the multinationals, the boards of trade, and middlemen, were convinced that they were failures.

5.

The Committee on Economic Development, mentioned earlier, is composed of 200 heads of large corporations, such as General Electric and Ford Motor Company, and economists from prestigious universities, such as Harvard and Stanford. The CED's Research and Policy Committee meets every other year to address various aspects of the U.S. economy, with the intention of providing "an aid to clearer understanding of the steps to be taken in achieving sustained growth of the American economy." And the purpose of sustained growth, they say, is to attain "high and secure standards of living for people in all walks of life . . ."

Need it be said that this is a purely quantitative goal, having nothing to do with quality of work, which in large part determines a person's quality of life? The committee presumes that a "high standard of living" can be purchased, and it is this "high standard of living," not equity or the absence of poverty, which it equates with the general welfare.

Indeed, some of the rhetoric of CED's 1962 report, "An Adaptive Program for Agriculture," sounds as if the committee is acting on behalf of farmers and the general public. But a close reading of the document shows that the committee's stated objective— reestablishment of a free market economy in agriculture—is at odds with its very recommendations, which depend upon a degree of government intervention never before contemplated in this country.

The basic premise for the report rests upon its claim

that "the industry"— agriculture—"is using too many resources." Indeed, the report routinely refers to farmers as "resources," another signal that people are regarded as commodities whose sole purpose is to consume other commodities. And even though the report notes that "net migration out of agriculture has been going on for 40 years, and at a rapid rate . . . . [it] has not been fast enough . . ." And this is because improved farm technology and increased capital had raised the living standards of the American people. Farmers, however, the report said, had not left agriculture to take advantage of it. (The CED never asked what would be the social, health, and environmental costs of moving more people into urban areas.)

However, if farmers could earn more outside of farming, they could consume more, which would mean more money for manufacturers. Besides this, farmers needed to be moved out of agriculture because the government, the report claimed, was spending too much on price supports and commodity purchases. The approach the committee recommended called "for action by the government working with the free market, not against it."

Out of the 200 members of the committee, only one noticed the irony. Fred C. Foy, chairman of Koppers Company, Inc., commented: "It is true that in some industries or areas of the economy labor and capital cannot earn as much income as they could if employed in some other use but who is CED to say that in this situation 'too many resources' are being used. In a free economy the owner of the labor or capital must be free to decide how he wishes to use them. It will always be true that some capital will earn less than others in the market place, but their earning less does not necessarily mean that they are being wasted or should be withdrawn."

The question for the CED was how farmers should be eliminated. The report noted three kinds of economic approaches to the question: laissez-faire, protectionist, and adaptive. The laissez-faire approach would remove all government programs at once, the consequence of which would be a precipitous and immediate decline in farm incomes. As a result, not only would the rate of exodus from farming increase, but as farmers retired, few would replace them. The laissez-faire process works best, CED said, when there are few resources needed to be shifted out of an industry and there is no "serious obstacle to the movement . . ."

The protectionist approach—which was the New Deal approach and still in operation during the time of the report—"usually requires government action." It affects an industry in various ways: by purchasing its product, by limiting its production or sales, and/or by subsidizing it. But the protectionist approach, the report claimed, "sacrifices the basic national interest in efficiency and growth . . ." (Note that efficiency and economic growth are equated with the nation's fundamental interest.)

Like the protectionist approach, the adaptive approach also utilizes government intervention, but the CED characterized this intervention as "positive government action to facilitate and promote the movement of labor and capital where they will be most productive and will earn the most income." Like the laissez-faire approach, the adaptive approach would shift farmers out of agriculture "but to do it more quickly and with less deep and protracted loss of income to the persons involved . . ."

The CED claimed that the history of U.S. farm policy showed the inadequacy of the protectionist approach, but did not substantiate its claim. It merely asserted that farm assistance put a burden upon the rest of the economy in the form of "high taxes, or high prices, or both." The country, claimed the report, "becomes increasingly resistive at the growing burden," which is "a basic national waste" of labor and capital. Besides, it said, we have no use for the great stockpile of commodities the program yields.

To protect farmers from the "temporary but sharp decline" in income that this approach would instigate, the CED recommended a five-year transition period. This would not mean an end to the movement out of agriculture, for with productivity continually increasing, there would continue to be an excess of farmers. This excess, however, "would be continuously moved out of agriculture." But, the report noted, "the transition we visualize will not bring itself about in a five year period. Action will be required to bring it about." This is truly an extraordinary statement, endorsed by all 200 members, with the exception of Mr. Foy. It stands in powerful opposition to the common notion that working Americans hold, namely the right to sell their labor, provided it be legal, to whomever they want, wherever they want. The CED is endorsing not a free market economy but centralized planning, every bit as centralized as the former Soviet system.

The CED transition program began with the recommendation that farm youths be induced to look for work outside of farming, and that schools provide them with vocational training. Those farmers who wanted to leave agriculture would also be given skills training, along with loans to help defray the cost of relocating to places of employment. This, the CED claimed, would be cheaper than the subsidy program.

But the CED knew that these "excess resources" would not leave agriculture until price supports on "cotton, wheat, rice and feed grains" were "allowed to reflect the estimated long run 'adjustment price' of these products." Adjustment prices would be somewhere between the support price and the free market price, but the drop to adjustment levels would be the "signal" for farmers to begin a rapid exodus.

To further hasten the demise of farmers, the CED recommended a Cropland Adjustment Program, by which farmers were to be induced to reconvert "at least 20 million acres of Western Plains and Mountain Region land from crop use to grass, as rapidly as possible." There would, however, be a Temporary Income Protection Program for wheat, rice, and cotton, to mitigate the harsh impact of the other measures. The amount of payments would decline over the five year period, at the end of which they would cease.

Third and last, the CED recommended a Temporary Soil Bank, which would pay farmers for taking crop lands out of production. The CED recommended that entire farms, rather than portions thereof, be removed.

It was possible, the report concluded, that these recommendations could result in 400,000 to 500,000 farmers leaving agriculture during each year of the five-year transition period.

The rigor of the program is summed up by CED member and University of Michigan economist Kenneth Boulding. "The only way I know to get toothpaste out of a tube," Boulding wrote, "is to squeeze, and the only way to get people out of agriculture is likewise to squeeze agriculture. If the toothpaste is thin, you don't have to squeeze very hard; on the other hand, if the toothpaste is thick, you have to put real pressure on it. If you don't get people out of agriculture easily, you're going to have to do farmers severe injustice in order to solve the problem of allocation."

Krebs noted that in 1964 U.S. Representative from North Carolina, Harold D. Cooley, Chairman of the House Agriculture Committee, said that between 1953 and 1962, "while all other segments of the economy had been booming, net income for agriculture had been $25 billion less than in the previous ten years." In that same period, USDA subsidies had dropped by $35 billion. Yet, Cooley noted, between 1953 to 1962, "Many farmers have turned against their own program—the program that prevailed during the years of our greatest era of prosperity. Why, and for what reason, I shall never understand."

CHAPTER FIVE
AGRICULTURE AND LAW

The United States is now engulfed in a battle between the forces of democracy and the power of money. For decades money has made ever greater inroads into the public domain, thanks to political campaign contributions in all three branches of government. Money has been able to buy politicians and hence write laws, while politically appointed bureaucrats sometimes fail to enforce existing laws. Money has unduly influenced state supreme court justices who, in many cases, interpret the law on behalf of money. A survey conducted in Texas, for example, showed that 83 percent of Texas voters, 79 percent of Texas lawyers, and almost half the justices on the Texas Supreme Court believe that campaign contributions unduly influence that court's decisions.

The public is losing faith in the judiciary. As U.S. Supreme Court Justice Anthony Kennedy noted, commenting on the Texas findings: ". . . . the law commands allegiance only if it commands respect. It commands respect only if the public thinks the judges are neutral. And when you have figures like that, the judicial system is in real trouble."

Loss of faith in democracy is rampant. According to the League of Women Voters, less than half of all registered voters —49.1 percent—turned out for the 1996 presidential election, while only 51 percent voted in the 2000 presidential election, and a mere 36.4 percent turned out for the 1998 U.S. Senate elections. When such large percentages of the population refuse to vote, and there is no third party strong enough to challenge entrenched interests, a sense of powerlessness pervades the country.

Meaningful protest through television, radio, magazines, or newspapers now seems unlikely since the most influential media are controlled by vertically integrated communications systems with horizontal links to other industries. This naturally exacerbates the public's sense of powerlessness, for what possibility is there of one vertically and horizontally integrated giant criticizing another, since the sins of one are the sins of all?

Has the struggle between money and democracy been decided? Are those advocates of decentralization, rural communities, and small farms fighting vainly against a Goliath that has already won? A March 26, 2000 front page headline in the *Des Moines Register* proclaimed: "A decline in farmers transforms Iowa." The article notes what many believe: "Family farming, as Iowa knew it in the 20th century, is irreversibly gone. A countryside dotted with 160-acre farms, each with its house and barn and vegetable garden, has all but disappeared."

The number of farms continues decreasing while the size of the average farm continues increasing. Not only farm wives have off-farm jobs: today 40 percent of Iowa's farmers earn off-farm income. Not unexpectedly, the number of farmers who work on contract for large producers has also grown. In 1987, agricultural economist Harold Breimeyer told Charles Kerault: "The main point is really your social and political image of what you think is better: whether one prefers a more egalitarian, democratic rural society in which each person on the land has some interest in his land and gets all the standing that goes with it or whether, instead, we go to a four- or five-layer-cake arrangement with the big owners in the cities, managers in the state capitals, the straw bosses in the county seats, and then two or three layers of workers on the land. That's where we are headed if we do not do anything to arrest the trend."

Fifteen years later the man's foresight has been proved 20-20. Neil Harl, nationally known agricultural economist at Iowa State University, says that the farmer's status as entrepreneur is endangered, that he is now "on the road to serfdom." As described earlier, this outcome has been the work of agricultural economists, top officials of the USDA, and leaders of American corporations going as far back as the Eisenhower administration.

The prosecution of Archer Daniels Midland for collusion with Japanese firms gave hope to a time to those who looked for our government to enforce its own laws. The federal anti-trust case against Microsoft further bolstered that hope, only to have it dashed when the decision against Microsoft was reversed by a higher court. Those examples coupled with the government's tolerance of oligopolies indicates that the crisis of law has passed and that the outcome in the battle between money and democracy has been decided in favor of the former.

2.

Agriculture, no less than any other function of the culture and economy, is controlled by vertically integrated corporate giants. According to some lawyers and legislators, it has been the lack of federal enforcement of anti-trust laws that has permitted a relative handful of corporations to dominate agriculture.

This problem of control by a few is called the problem of concentration. One of the best known opponents of corporate concentration in agriculture is William Heffernan, a rural sociologist at the University of Missouri. Heffernan uses an apt analogy to explain concentration. "The [agricultural] system," he wrote, "is like an hourglass: commodities produced by thousands of farmers pass through a few large firms to millions of consumers. As a result the market is no longer competitive and information about the market is no longer accessible."

Concentration in agriculture was the focus of an August 1999 listening session in Cedar Rapids for Iowa Senators Harkin and Grassley, Minnesota Senator Paul Wellstone, Assistant Secretary of Agriculture Mike Dunn, and the head of the anti-trust division of the Department of Justice, Joel Klein.

Approximately 350 farmers and family farm advocates arrived to tell the politicians and bureaucrats where they stood on this matter.

The consensus among the experts, including agricultural economist Neil Harl of Iowa State University and Jon Lauck, formerly of the University of Minnesota Law School, was that concentration needs to be attacked and that existing anti-trust legislation is sufficient to address the problem. What is and has been lacking is vigorous federal enforcement. Speaking over a telephone speaker, Harl told the assembled group that "Section 1 of the Sherman Act"—which prohibits collusion among firms—"needs to be looked at."

Everyone in the audience at Cedar Rapids knew that the issue, as Heffernan wrote, was control. The control that a handful of corporate alliances exert in all phases of agriculture, from seed production to contract farming, processing, storage, and packing. Even shipping. Heffernan identifies three major alliances—Novartis/Archer Daniels Midland, Cargill/Monsanto, and ConAgra.

The big grain buyers, such as ConAgra, own their own fleets of tows and barges. They even own freighters that ship their grain from various U.S. ports to destinations worldwide. Each major grain buyer has alliances or contractual agreements with other vertically integrated firms in other areas of the food system, as noted above. Cargill, for example, has partnering arrangements with Monsanto. The industry journal, *Grainnet*, reported that the Cargill/Monsanto cluster controls "40% of all corn exports, 33% of all soybean exports, and 20% of all wheat exports." Cargill ranks in the top four firms producing animal feed, and in feeding and processing cattle. ConAgra, Heffernan reported, owns "Armour, Monfort, Swift, Butterball, Healthy Choice, Peter Pan Peanut Butter, Hunt's and many more." Further, within each of these clusters is a joint venture or alliance with a biotechnology firm.

"For well over a decade," Heffernan wrote in "Consolidation of the Food and Agriculture System," "several of us at the University of Missouri have been reporting the concentration ratios of the largest four processors of most of the major commodities produced in the Midwest . . . . We focus on the largest four processing firms because the economic literature in the mid-1980's indicated there was general agreement that if four firms had 40 percent of the market, that market was no longer competitive." Heffernan says his data indicates that 40 percent mark has been reached.

Over time it has become increasingly difficult for him to collect data on agreements, mergers, and other deals. "Trade journals have come under pressure to not publish some of this information and government agencies often say that to reveal the proportion of a market controlled by a single firm in such a concentrated market is revealing proprietary information." In other words, our government is promoting the interest of big business, whether or not that interest works on the public's behalf. Today Heffernan and others cannot agree on exactly what percentage the four largest firms control, and the fact that he and others differ by six percentage points makes him uneasy. Furthermore, data collection has become increasingly difficult with the Asian and other foreign ventures in which the alliances are engaged. Their control of the food system is not merely national, but global.

How many clusters will eventually emerge? Heffernan is predicting four or five "because the number of clusters will be heavily influenced by the number of firms who have access to the intellectual property rights [to the gene pool]." The biotech research leaders—Monsanto, Novartis, Dow Chemical, and Dupont—are already tied into existing clusters. Given the enormous expense of research and the patents they have been granted for biotech products, Neil Harl thinks it unlikely that any new cluster will emerge.

The Cedar Rapids convocation did not focus on biotechnology, but on other facets of the food system that are reducing competition. One of the immediate concerns for the Cedar Rapids group was the approval by the Department of Justice of Cargill's acquisition of Continental's grain operations. Joel Klein, the DOJ's head of anti-trust, defended the decision on the grounds that it called for Cargill/Continental to divest itself of elevators in Stockton, California; Beaumont,

Texas; and Seattle, Washington.

In "Agrarian Antitrust" Jon Lauck noted that the DOJ barred acquisition where farmers had to choose between Cargill or Continental. "DOJ quite obviously stopped Cargill's acquisition of Continental's facilities in areas such as the Pacific Northwest where the acquisition would leave only one major grain buyer. In short, DOJ prevented duopoly from devolving into monopoly." But, as Lauck noted, "it failed to recognize the great potential for cooperation and collusion in heavily concentrated markets. It failed to recognize the unique bargaining disparity between disorganized farmers and large-scale agribusiness firms." Finally, Lauck noted that the DOJ's decision "failed to respect a series of statutes passed by Congress and state legislatures concerned about the concentration problem in agricultural markets." This failure has brought challenges to the ruling from several states attorneys general.

3.

As stated earlier, each of the food chain clusters—Cargill/Monsanto, ConAgra, and Novartis/Archer Daniels Midland—has within its alliance a biotechnology research firm. For Harl: "Increased concentration is also leading to control by a few firms over the major processes by which genetic manipulation occurs, thus enabling those controlling the technologies to block use by other firms."

Harl and Heffernan's concerns with patenting genes and transgenic species focuses on the problem of concentration, claiming that the system will force out competition. Harl believes that germ plasm should be public domain. He is arguing in the tradition of natural law, expressed in the *Institutes of Justinian*: "Thus, the following things are by natural law common to all—the air, running water, the sea, and consequently the sea-shore."

But the exploitation of germ plasm, whether public domain or not, is unconscionable. Of all legal issues in the twentieth century, genetic engineering, for its inevitable consequences, may loom largest in years ahead. Biotechnology received its impetus from the U.S. Supreme Court decision of 1980 (Diamond v. Chakrabarty), which granted patent rights for a novel bacterium. Arguing by precedent, it follows that all life forms can be patented. Indeed, since that 1980 decision, the U.S. Patent Office has granted patents for plant, animal, and human genes, and to transgenic plants and animals. Plants, such as tomatoes, soybeans, and corn, are now genetically engineered, and they and their by-products are on our supermarket shelves. Animals, such as hogs and chickens, have been successfully engineered, but are not yet commercially available.

Assuming nature behaves as researchers think it does, human gene research will find its first application in the cure of genetically caused diseases and will advance to the genetic engineering of qualities such as hair and eye color, then proceed to more fundamental qualities such as intelligence. The scientific possibility or likelihood of this engineering feat will be discussed later, but since people now routinely sell their sperm and eggs, there is nothing to prevent joint ventures between companies which hold patents on human genes from attempting to engineer human beings from start to finish. Legally, it is possible. Few people will object to finding a genetically engineered cure for an inherited disease, and once legal precedent for human engineering has been set (assuming it can be accomplished), nothing will prevent more fundamental tinkering. As with any highly questionable procedure, the public will first be won over by benevolent applications, but in so doing it will find the lines between gene therapy and bioengineered enhancements blurred.

Our scientific and medical techniques have led us to expect that nothing need be left as it came from Nature, and in the absence of the sacred, Nature herself has become little else but a field for human manipulation. As we have become accustomed to intrusive exercises of all sorts, including cloning, we have become desensitized to our manipulations and lose our moral orientation. As Richard Weaver argued in *Ideas*

Have Consequences, establishing the fact of moral decline is the most pressing task of our time. Its "most serious obstacle is that people traveling this downward path develop an insensibility which increases with their degradation. Loss is perceived most clearly at the beginning; after habit becomes implanted, one beholds the anomalous situation of apathy mounting as the moral crisis deepens." This, he says, is why the scholastics "were extremely agitated over questions which seem to us today without point or relevance. If one goes on, the monitory voices fade out, and it is not impossible for him to reach a state in which his entire moral orientation is lost."

It should be obvious that the U.S. Supreme Court's 1980 decision allowing the patenting of life forms can have no weight. The court simply does not have the authority to assign anyone or any institution the rights to genes or life forms. They are not subject to ownership. The court has arrogated to itself powers it simply does not have, and the inability to see this fact is one more instance of modern man's loss of moral orientation.

4.

The convocation at Cedar Rapids had other interests, one being the need for mandatory price reporting for cattle, hog, and sheep sales. Since four packers control 82 percent of cattle slaughtered, and six control 60 percent of all hogs slaughtered, independent producers were saying that lack of reporting was concealing the fact that larger producers like Decoster, Smithfield, and Iowa Select were receiving special premiums. Indeed, a farm woman on one of the Cedar Rapids panels said she and her husband gave up contract hog farming for a large producer when they grew ashamed of how much more they were earning than their neighbors, who were independent hog farmers.

Others present, including Iowa State Representative Jack Kibbie, speaking on behalf of six state representatives from six Midwestern states, said the problem could be dealt with through the Packers and Stockyards Act. One anonymous handout at the Cedar Rapids session asked: "Why hasn't the Packers & Stockyards Act passed in 1921 been rigidly enforced? This strongly worded law was designed to ensure free, open competitive and fair markets to all livestock producers." It went on to say that the act needed to be enforced for several reasons, first and foremost because we "must not become known as a lawless nation, who [sic] has laws but does not enforce them!"

The Packers & Stockyards Act mandated that all sellers be treated equally and fairly. It provided protection against price discrimination and stipulated that "all prices and transactions shall be recorded on a daily basis, including true ownership of business." And finally, the act granted the Secretary of Agriculture the power to "act and call for whatever records he deems necessary" and to fund such investigation. Farmers had been outraged over price secrecy for some time, and small farm advocates had been writing about it. Even the American Farm Bureau, traditionally on the side of agribusiness, lobbied Congress for mandatory reporting.

Finally, in October of 1999, the small farmer found consolation in the Livestock Mandatory Reporting Act, which required processors to list prices paid, including any special premiums. But by that time the damage had been done. The markets were glutted with everything from corn to wheat to hogs, and the independent producer was still buying seed and fertilizer and chemicals from one of the three clusters of firms. And if the independent producer was paying X dollars for so much feed, wasn't his giant competitor getting a better deal for buying in volume?

5.

On the state level, the connection between money and law is well illustrated in both Iowa and North Carolina, which are respectively the number one and number two hog producing states in the nation. North Carolina is the home of Wendell Murphy, founder and former president of Murphy Family Farms, who was described by *Forbes Magazine* as "the Ray Kroc of hogs." As the father of factory hog production, Murphy turned his family hog farm into the country's largest hog producer. Along the way he became a North Carolina state senator, helping to craft legislation which nurtured large scale hog production.

In 1995, Murphy was the subject of a five-part Pulitzer Prize winning report, "Boss Hog," in the Raleigh, North Carolina newspaper, *The News & Observer*, which reported that during his ten years in the North Carolina General Assembly, Murphy either sponsored or helped pass legislation that exempted hog factories from local zoning ordinances, from stiff environmental regulations, and from state sales tax.

Murphy's friend, Sen. Harold Hardison, sponsored 1987 legislation which exempted hog and poultry operations from state sales tax, which the state later calculated cost North Carolina almost $1 million for the period prior to 1990. Hardison admitted getting a "sizable [campaign] contribution" from Murphy, but did not recall the amount.

A year later, in 1988, Hardison made a bid for the Democratic nomination for Lieutenant Governor and, according to *The News & Observer*, received a $100,000 contribution from Murphy and another $100,000 from Marvin Johnson, president of a giant turkey processing firm. The legal limit for campaign contributions in North Carolina is $4,000. *The News & Observer* reported that: "Those and other illegal contributions to Hardison were uncovered by the State Bureau of Investigation." It also reported that Wake County district attorney Colon W. Willoughby "could not prosecute Murphy or Johnson for election law violations because the two year statute of limitations had expired."

In 1991 Murphy co-sponsored a bill that exempted factory farms from county zoning jurisdiction. That same year he sponsored an amendment that would have made it illegal for the state to fine hog factories or any other farm operation from discharge of manure into streams or water supplies.

Between 1990 and 1995, reported *The News & Observer*, "Murphy, members of his family and his executives have contributed about $150,000 . . . to local candidates, legislators, members of Congress and others, right on up to Gov. Jim Hunt."

The News & Observer reported that for a two-year period the Murphy company and family received an average of fifty calls a week from the governor's office, legislators, and state agencies. One of the legislators was a Murphy neighbor and chairman of the Senate agriculture committee.

The lack of regulations in North Carolina meant that hundreds of earthen lagoons dotted the eastern portion of the state. Spills and leakage became normal occurrences. A report issued by the minority staff of the United States Senate Committee on Agriculture, "Animal Waste Pollution in America: An Emerging National Problem," noted that "In 1995 in North Carolina: 35 million gallons of animal waste spilled into the state's waterways. Kills of 10 million fish were attributed to animal waste pollution. Animal waste pollution closed over 360,000 acres of coastal wetlands to shellfish harvesting."

A toxic microbe, Pfiesteria piscicida, developed in North Carolina streams as a direct result of manure runoff. The Senate minority committee report stated: "In 1997, an estimated 450,000 fish were killed in North Carolina by the toxic microbe Pfiesteria piscicida, whose increased presence in estuarine waters is linked to excess nutrients from animal waste and farm runoff." It also noted: "Symptoms reported among people with close exposure to Pfiesteria in its toxic form include memory loss, respiratory problems and skin rashes."

Despite these problems—despite the death of several rivers, the fish kills, the irreparable damage done to people—the hog industry made no effort to reduce or eliminate runoff. Waste was still routinely sprayed onto fields adjacent to rivers.

Then in 1999 came Hurricane Floyd, sweeping over eastern North Carolina, flooding lagoons and carrying hogs and hog waste across the eastern lowlands, flooding towns like New Bern, spreading its toxic liquids and particulate matter everywhere. Even after Smithfield Foods, the country's largest hog producer, agreed to a settlement of $65 million with the State of North Carolina, the industry's hogs and its 2,500 lagoons remain, waiting for another hurricane.

Of course political campaign contributions help explain a lot. In explaining why he made a donation to Republican Jesse Helms, businessman and Democrat John Belks said: "I do a lot of business up in Washington. If you don't have contacts, you don't have any influence. It's pretty obvious that if you're going to be active in politics, you've got to have friends on both sides." As another contributor explained: "I'm in business, and a lot of times things come up when I'd like to have some communication with whoever sits in the governor's house. If I gave nothing, I would feel like I didn't have the right to express my views or communicate with them." In 1998, Democracy South, a North Carolina non-profit that gathered the above quotes, wrote: "Through both legal and illegal means, North Carolina businesses have spent more than $7 million to influence the outcome of [state and national] elections in the past three years."

6.

In 1997, Nebraska citizens began mobilizing against hog confinement operations. In Nebraska, farmers toured the state, speaking passionately and eloquently of how the quality of their lives had been hurt by the vast hog confinement operations that had been erected near them. They spoke about the unbearable stench these factories generate. Hog gas, they said, moves in a mass and is not easily dispersed by winds. If it drifts into your house and the wind stops, you will be choked with it. It permeates your clothes and stays with you. Hog manure from these huge operations fouls the water and makes it undrinkable. Hog confinements lower property values and divide the neighborhood, people lining up on both sides of the issue, neighbor against neighbor. The farmers accused county supervisors of being bought off. They were warning people across Nebraska to organize before it was too late.

A year later, factory hog farms became an issue for Iowa. Rural neighbors suddenly found themselves angrily divided. Some were supporting contract farmers' right to do whatever they wanted on their property, especially when they needed some way to stay in farming. Others would say the stench and property devaluation, along with the potential for water contamination, would be too great to ignore and that the farmer had to be challenged. But those who challenged factory farms wondered if their court petitions would be accepted.

All this happened because Iowa, despite North Carolina's massive fish kills and documented contamination of ground and surface water, worried that its southern competitor would become first in hog production. To encourage large scale operations, the Iowa legislature passed a law protecting farming operations from litigation by neighbors who might sue for property devaluation, odor, or ground water contamination.

No matter that the law stripped Iowans of an American's most fundamental right: redress of grievance before the law. To make matters worse, the legislature had already stripped counties of the power to determine their own policy on factory farms. If a county did not want them, too bad. If a county wanted to exercise control on siting, too bad. Most Iowans were beginning to understand where they stood in the eyes of their political bosses. It became common to hear people acknowledge that it was easier for money to control one set of legislators in Des Moines than ninety-nine sets of county supervisors across the state. Local determination, which the Republicans had touted in their race to dismantle federal programs, suddenly meant nothing.

The tool by which then Governor Terry Branstad and the legislature successfully wooed industrial hog producers to the state was a 1995 bill set forth in Iowa Code Section 657.11 that granted nuisance suit protection for corporate hog factories, which the code refers to as "animal feeding operations."

Section 36 of Iowa Code 657.11 deals with animal feeding operations. The section begins: "The purpose of this section is to protect animal agricultural producers who manage their operation according to state and federal requirements from the costs of defending nuisance suits, which negatively impact upon Iowa's competitive economic position and discourage persons from entering into animal agricultural production." It is clear from the outset that the statute's purpose is to encourage hog confinement operations. "This section is intended to promote the expansion of animal agriculture in this state by protecting persons engaged in the care and feeding of animals. The general assembly has balanced all competing interests and declares its intent to protect and preserve animal agricultural production operations."

The statute's next paragraph declares that once "a person has received all permits required . . . for an animal feeding operation . . . there shall be a rebuttable presumption that an animal feeding operation is not a public or private nuisance . . . and that the animal feeding operation does not unreasonably and continuously interfere with another person's comfortable use and enjoyment of the person's life or property under any cause of action."

In order to win a nuisance suit one has to prove that the "animal feeding operation unreasonably and continuously interferes with another person's comfortable use and enjoyment of the person's life or property" and that "the injury or damage is proximately caused by the negligent operation of the animal feeding operation."

So first, a plaintiff had to prove that the hog factory unreasonably interfered with his enjoyment of life and property. And what is unreasonable? If farmer Jones finds that the only way he can keep raising hogs is to sign a contract and become indentured to Smithfield or Iowa Select or Carroll Foods, the law must balance his right to farm with his neighbors' right to live without their property being devalued by his operation, and without the sickening stench of hog feces and the possibility of ground and surface water contamination. An agriculturally leaning Iowa court might well conclude that the stench, water contamination, and property devaluation are not unreasonable in light of farmer Jones's need to earn a living. After all, Iowa's legislators and governor felt that corporate interests far outweighed the health and welfare of Iowa's citizens. Indeed, the court might conclude that it was unreasonable of us to expect farmer Jones, in view of his need, not to contract with the XYZ corporation.

But supposing that the court did conclude that the nuisance was unreasonable. Still, the unlucky plaintiff would have to prove that the nuisance was continuous. And what exactly does that mean? That the stench is present twenty-four hours a day? Does twenty-three hours count? Nineteen? Ten? Does "continual nuisance," applied to drinking water, mean that your drinking water, which is now contaminated, will always be contaminated? The injured party begins wondering just what he is going to have to prove in court. He begins to wonder if he has a ghost of a chance. Moreover, even if the injured party lives with the stench twenty-four hours a day, seven days a week, fifty-two weeks a year, and even if his drinking water is contaminated with a high level of nitrates, still he would have to prove that this was due to farmer Jones's negligence.

Hog stench can't be laid to negligence, because hog feces naturally stink. As for water contamination, suppose Jones stored his slurry in an earthen lagoon and it leaked. How would you prove negligence? Since earthen lagoons are permitted under state law, and since earthen lagoons leak, leakage can be blamed on materials, not negligence. The piece de resistance of the legislation was the paragraph that declared that if the plaintiff lost, he would be liable to the defendant "for all costs and expenses incurred in the defense of the action, if the court determines that a claim is frivolous."

Several years before the statue was ever challenged in court, lawyers had said that if it were ever offered in defense that it would be declared unconstitutional. The Iowa Civil Liberties Foundation had a brief prepared and was waiting for a defendant—Carroll Foods or Smithfield, or whomever—to offer the statute in defense. Finally, two defendents did, and at both trials judges ruled the code unconstiututional. One defendent filed an appeal. This was surprising, since the code, as one Iowa attorney noted, was worth more to the pork industry as a boogey man on the books than in court. The final decision, without doubt, will have a significant impact on the mega-hog producers, who are already considering Brazil and Poland as the next centers of concentration.

7.

When Humboldt County supervisors tried to exercise home rule prerogatives to control the siting and size of hog confinement operations, they learned a lesson that resounded across Iowa. In trying to prevent manure runoff from hog factories polluting streams and aquifers, the supervisors invoked two ordinances. One required a county building permit, which would control siting, and the other required the factory owner to be insured against environmental damage caused by spills, runoff, and abandoned property.

The supervisors were particularly concerned with the potential damage posed by agricultural drainage wells. In the early 1900s the U.S. government began allowing farmers to drill down to aquifers, and consequently surface water since that time has flowed directly into them. The county has 200 registered agricultural drainage wells and estimates it has between 350 and 400 unregistered wells. "Most of these," said the county's environmental protection officer Doug Wood, "are direct conduits into the aquifers. We wanted to put these things [animal factories] where they didn't endanger agricultural and drinking wells or our rivers and streams."

Lloyd Godell and his three sons wanted to erect a hog factory in Humboldt County, but every site they picked was vetoed by the supervisors as environmentally unsuitable. The Godells, joined by twenty contract feeders and seven independent hog producers (all supported by the Iowa Pork Producers), took Humboldt County to district court. The court ruled that the county, under home rule protection, had the power to control the size and siting of agricultural operations. But the plaintiffs appealed the case to the Iowa Supreme Court which ruled against the county on March 5, 1998, citing two laws. One was an agricultural zoning law, enacted in the fifties, which has been fine tuned to include anything related to agriculture. The other was Iowa Code 657.11.

In its decision, the court addressed the effect of the 1978 Constitutional Amendment granting "home rule" to counties and municipalities. Under that amendment, counties have the power "to determine their local affairs and government" but only to the extent those determinations are "not inconsistent with the laws of the general assembly."

The court found that four of the Humboldt County ordinances were in conflict with state law. Specifically, it declared that Ordinance 22, which required a construction permit for confined animal feeding operations (CAFOs), conflicted with a similar state statute; that Ordinance 23, which established security requirements for CAFOs, conflicted with the same state statute; that Ordinance 24, which set guidelines for protecting groundwater, conflicted with a statute giving the Department of Natural Resources sole authority over manure disposal; and that Ordinance 25, which would have established air quality emission standards for CAFOs, conflicted with Iowa Code 657.11.

Years before, in 1976, the Iowa Supreme Court had upheld Humboldt County's decision to prevent a chicken factory from locating in the county. "At that time," said Doug Wood, "home rule was okay. But under pressure from the National Pork Producers, any ag related business is now protected."

The combination of nuisance suit protection and the Humboldt County case caused a wave of anger and a sense of powerlessness to sweep the state, as the *Des Moines Register* discovered in so many angry letters on the hog confinement issue that it eventually told its readers to cease sending them. The message, however, from the legislature and supreme court was clear: not all citizens or corporations are equal in the eyes of the law.

8.

In his January 1998 annual State of the Judiciary address, Iowa Supreme Court Chief Justice Arthur McGiverin expressed his fear that judicial independence was under attack. "The word 'independence,' he wrote, "refers to independence in decision-making. Judicial independence means upholding the law without fear of the consequence of political retribution."

To make his point, McGiverin cited a murder trial in which a jury convicted a woman of second-degree murder, a decision which prompted public outrage. The judge, apparently fearful of losing re-election, decided that "he had erred by not allowing the jury to consider the possibility of a manslaughter conviction."

"Judges must not test the winds of public opinion before entering a decision," wrote McGiverin. "They are bound by their oath of office to render decisions based on the constitution and statutory law." This is what we expect of our judicial system. It is what any republic expects of its constitution and legal code: protection from tyranny and mob rule, which is to say protection from the vagaries of arbitrary will. As McGiverin wrote: "Without the rule of law, our legal system would be unstable and unpredictable. . . . Judicial independence ensures that judges defy current popular opinion in favor of the long-standing principles established in our constitution. Judicial independence ensures social order and stability. . ."

There are, he claimed, too many instances of groups attempting to affect judicial decisions who are thereby undermining judicial independence. Among his examples were some "of our district court judges [who] have come under attack for their decisions in hog lot cases. These attacks are not based on the soundness of the judges' legal analyses but on the critics' unhappiness with the outcome. I've been told that anti-hog lot organizers, who came to the state house in November to attend the Supreme Court hearing of a hog lot case, reminded their followers to vote against the justices in the next retention election if the court didn't rule in favor of the group's cause."

McGiverin concluded: "For judges, criticism comes with the territory and we are entirely accustomed to it. However, the kind of efforts I have just described threaten the integrity of our nation's justice system."

McGiverin's critique is erroneous because it is inconsistent. On the one hand, he accepts the presence of criticism and insists that judges remain detached from it. On the other hand, when judicial decisions have been influenced by public pressure, he faults the public, not the judges, for the latter's lack of integrity. But the lack of integrity of some judges is an extremely serious matter, for it brings the judicial system into disrepute. More than any one else we expect integrity and honor from judges, who stand as a symbol of law. To repeat the words of U.S. Supreme Court Justice Arthur Kennedy: ".... the law commands allegiance only if it commands respect."

CHAPTER SIX
THE TRIUMPH OF TECHNIQUE

The fact of the death of nature is nowhere proclaimed so loudly as in genetic engineering, whose greatest triumph is the creation and manufacture of new species whereby results are obtained, not by crossbreeding, but by intervention across species boundaries. And of all the scientific techniques developed in this century, genetic engineering is one of the most astonishing, perhaps the most audacious. In this country we hear little public opposition to it, while in Europe and Asia public outcry has banned imports of genetically engineered American crops and livestock. America's public silence, I think, is owing to the fact that relatively few Americans know what genetic engineering involves, and the extent to which it is practiced. Indeed, outside of rural areas, how many Americans know that most corn and soybeans are now genetically engineered, and that approximately 70 percent of all processed foods contain bioengineered ingredients? Companies that stand to make a profit from these products have, with the Food and Drug Administration's complicity, kept them free of labels that would alert buyers to their contents. In addition, the media covers the subject only occasionally, and then the stories appear to be limited to periodicals or programs addressed to a relatively small sector of the population. On the other hand, there are no lack of books and journal articles on the subject.

To understand the enormous implications of genetic engineering we must explore areas generally untouched in writings on the subject. Since the discussion is for the most part directed by rationalists—those who believe that virtually any human activity can be controlled and directed by reason—the arguments for genetic engineering are couched in the rationalists' own terms, which means that larger issues on the nature and origin and value of life are ignored. What we need to do, first of all, is to explore the rationalists' presuppositions and the question of how genetic engineering came to be practiced in the first place.

Genetic engineering is the most sophisticated aspect of our highly industrialized contemporary agriculture. When we think of industrialization we usually think of machinery, and in farming we think of tractors, spray rigs, combines, milking machines, and so on, all obvious labor saving devices. But industrialization involves standardization and the maximization of efficiency, both of which are at the heart of today's agriculture. Indeed, they are at the heart of the doctrine of progress. Now since the doctrine of progress proclaimed that human knowledge would continue to grow endlessly into the indefinite future, providing us with ever more material goods and prosperity, it encouraged the ongoing development of scientific techniques and tools. After all, these made scientific advancement, and therefore progress, possible. In our time this translated into the mantra, Newer is Better.

As far as thousands of scientists and technicians are concerned, genetic engineering is not only an instrument of progress, it is probably the greatest and most beneficial technical tool ever developed in the history of mankind. This point is made repeatedly in *The Thread of Life*, written by Susan Aldridge and published by Cambridge University Press. So glowing is Aldridge's assessment of genetic engineering that her work might as well have been written by a public relations counsel for Monsanto. Far from seeing any potential harm in this technique, Aldridge writes that "its potential comes from humans, rather than the blind forces of evolution, being at the controls." In other words, humans can do a better job of directing Nature than Nature herself!

2.

Like the powers summoned by the sorcerer's apprentice in Goethe's fable, our technologies are out of control. Unlike the apprentice, who had his master, we have no one who can restrain them. We are left with our technologies and machines—our techniques—running our lives. What domains of human expression and experience remain exempt from their deadening hand?

"No social, human, or spiritual fact," wrote French sociologist Jacques Ellul, "is so important as the fact of technique in the modern world. And yet no subject is so little understood."

Let us distinguish between techniques and technique itself. Techniques are tools and methodologies; technique proceeds them. It is an idea. Techniques are its offspring.

Technique is never satisfied with its children. It must always create new techniques to supersede the old. Its criterion is efficiency.

Technique, for example, is not satisfied with a human hand and a pen. It demands the typewriter, the word processor, the photocopier, the fax machine. The world will never again produce illuminated manuscripts.

Once the machine with its regularity became regarded as the model for all human activity, man began to transform his institutions and activities into machines.

Technique then expanded into every sphere of life. At first external to man, technique progressively absorbed him. Now it has entered every facet of his life.

Technique has even absorbed politics, thanks to public relations and advertising. Rehearsal is an integral part of contemporary political technique.

Technique creates efficiency and order. It leaves no room for chance, for human impulse, for idiosyncratic behavior. It wishes to expunge the human presence from the world.

Technique makes mass production possible. It likes uniformity, in products and people. It is opposed to handcrafted goods and indigenous cultures. Technique seeks to impose itself on all peoples world-wide.

Technique demands efficiency. A sterile woman and her husband are inefficient. Technique supplies them with in vitro fertilization.

Technique suggests the genetic engineering of man, plants, and animals. Technique does not want to leave the human prospect to chance. Technique proceeds from predetermined ends. Once an end is determined techniques will be developed to facilitate its accomplishment. Technique is the totality of means used to accomplish a predetermined end.

Technique began as a means to ease man's burden; now man exists to sustain technique.

3.

To achieve this ascendance, technique needed philosophical materialism, the doctrine which maintains that all that exists is composed of matter. So long as society was religious, reality was said to belong to the immaterial, transcendent world. So long as most of society still acknowledged a transcendent realm, they acknowledged an Origin and Center, and traditional man defined himself in relation to it. But with the development of scientific techniques in the seventeenth century, and the practical benefits they bestowed, the weight of reality shifted to the material world. Eventually, materialism was used to discredit the transcendent, which was banished from the public realm.

This left the field free to experts who could think objectively and scientifically. Since the Enlightenment, modern bureaucracies, armies, and corporations have been designed to act by the dictates of reason, that is, according to method or technique. Their designers conceived of the world as divided into separate and discrete domains which could be manipulated with precision. Hence, a class of managers, bureaucrats, scientists, and professional army officers arose, each of which was trained to manage and direct affairs in its domain. The problem that consistently arises is that such narrow specialists, trained as rationalists, are often incapable of meeting unexpected situations or predicting the outcome of their actions.

For example, reason dictates that heavy chemical application on farmland is the most efficient way to eliminate weeds and insect pests, and thus maximize crop production.

In fact, tests are now showing that organic methods are more productive than chemical methods, but this does not fit with the rationalists' preconceptions. So the rationalist must deny the accuracy of the results or validity of the testing method. Further, because the agricultural expert's method is focused on efficiency and profit, it cannot and will not take into account health and environmental factors. Thus, in the case of the dead zone in the Gulf of Mexico, the experts, including the American Farm Bureau, have: a) denied its severity and its impact on the fishing industry, b) maintained that there has been no demonstrated link between the dead zone and the nitrogen in chemical fertilizers and hog manure, and c) pointed out that the American farmer must be allowed to make a living, despite occasional and very minor harm to the environment. Of course, much of what experts say in defense of current agricultural methods is simply because huge amounts of money are invested in them. Nevertheless, if the rationalists did not believe in their methods, they would not make their investments an all or nothing proposition. In other words, if the experts determine that chemical farming is the best method, then all farming must be chemical intensive.

It has taken years of environmental opposition to chemical farming for agribusiness to devise an alternative, but the solution is like throwing gasoline on a fire: Monsanto, Pioneer Seeds and other firms have engineered pesticide resistance into crops! Alternatively, Monsanto's Roundup Ready Corn is resistant to Roundup, its pesticide. As Agronomist Dennis Keeney commented in a letter: "some weeds will develop resistance and sooner or later the technology will be ineffective. This is the story of most herbicides." Furthermore, as with chemicals, there is no thought for possible consequences to the environment and human health.

So great are potential profits, and so wedded is agribusiness to technique, that it assumes that problems in one box, in this case crops and animals, will not affect matters in another. Thus the advocates of genetic engineering cannot conceive of the possibility that introducing the anti-freeze gene from the flounder into tomato DNA through a virus might eventually prove troublesome, if not lethal, to other species. In 2000, a spokeswoman for the USDA assured the Senate Committee on Agriculture that genetically engineered foods do not pose a health risk, therefore do not need labeling. But as a spokesman for the Consumer's Union retorted, Who is to say whether twenty years down the road problems will not arise? The USDA's assumption, he said, "is an expression of faith, not of knowledge." And he is correct.

There is no way to assure the public that transgenic foods are safe, because there is no reliable way to test for dangerous substances. Therefore Monsanto and its allies, including the USDA (which shares patents with universities and corporations on a number of genetically modified plants and animals), have persuaded federal regulatory agencies that genetic engineering is merely an extension of conventional breeding. But it is not simply an extension of the species modification that humankind has practiced for millennia. Selectively breeding animals of the same species is not equivalent to human intervention across species.

But faith in science and technology is part of what drives our economy. Greed, of course, provides even greater incentive, while science and technology provide the tools to design the products and the rationalizations the elites deliver to the public. After all, experts know best, and if a battery of scientists approve a new technique or product, what can we say in reply? Especially when an increasing number of communications corporations are being vertically integrated into larger corporate structures.

4.

In 1954 Ellul wrote: "Technique, or rather techniques, appear as the motive force and the foundation of the economy. Without them, there is no economy." This fact is illustrated in the recent history of agriculture, in which the continued worldwide growth in the sale of farm equipment and supplies, including seeds, fertilizers, herbicides, and pesticides, as well as the sales of processed and unprocessed meats and vegetables, is based on the continuous development of new products and techniques. For years agribusiness spent millions of dollars in developing newer and newer strains of hybrid seeds. Now the seeds are genetically engineered. Fertilizers, herbicides, and pesticides are likewise subject to improvement. Food is processed by newer methods, or packaged in different ways. Meat is irradiated. Dairy cows are given BST to stimulate greater milk production. And so on. These techniques and new products (which themselves are often the result of new techniques) constitute the essence of the today's industrialized agriculture, and it is their proliferation which has been a major contributing factor in the demise of the small and medium size farm, and consequently of rural America itself.

Rural America has declined because small towns depend upon the prosperity of numerous small farms, and upon farmers spending locally. A precipitous decrease in the number of farms, such as we have seen in the last twenty years, has a catastrophic effect on rural villages. But the small and medium size farms are failing not only because middlemen grab most of the profit from the sale of crops and livestock, but because the farmer, like the businessman, has been persuaded to adopt a constant stream of new technologies, which come at a high cost.

The essence of contemporary business is its demand for profit: this year's profit can never fall below last year's; the present quarter's earnings can never fall below the previous quarter's. Simply maintaining the last quarter's earnings is unacceptable, but to maintain ever increasing profit business has come to depend upon innovation.

Technique has been driving agriculture for at least the last fifty years, and its most active offspring, genetic engineering, is a booming business. Investors sees billions in potential profit from the creation of new plants and animals.

Genetic engineering, as the phrase implies, involves the manipulation of genes, protein molecules within the nucleus of an organism's every cell, which carry the organism's inherited characteristics. They are found on strands of deoxyribonucleic acid, known as DNA, strands of which are coiled together into chromosomes. The DNA in every cell of a plant or animal is identical with every other strand of DNA in that plant or animal, and is formed of the molecular combination of a sugar, a phosphate, and a base. It appears in dual strands, winding about, in the form of a double helix. Each strand consists of two spirals—composed of sugar and phosphate—connected by four bases—adenine (A), guanine (G), thymine (T), and cytosine (C). Each spiral has, in unpredictable order, these four bases projecting outward, at an angle. And each base is met by another base projecting from the other strand, but A can only be met by T, and G can only be met by C. Each base pair is joined by a weak hydrogen bond. The resulting molecule resembles an extremely long, winding staircase of 3 billion steps. It is the order of the base sequences which give a DNA molecule its particular genetic information.

What genetic engineering does is to find the sequence of bases for a particular gene, snip it out (sometimes using an enzyme), and insert it into the receiving DNA. One of the greatest fears aroused by genetic engineering comes from the fact that genes are often integrated into a host through a vector (carrier), which is either a retrovirus, whose pathogenic potential has been neutralized, or a plasmid. Plasmids are rings of DNA which are antibiotic resistant. The one most commonly used in creating transgenic plants is a plasmid carried by a bacterium (Agrobacterium tumefaciens). (The use of plasmids and viruses create a number of hazards that will be discussed below.) A second method is to attach the gene onto the surface of a tiny golden bead and "shoot" it into the host cell. The third method is microinjection, in which the new gene is mixed in a solution and injected into the host.

Technicians never know where a gene, inserted by any of these methods, will attach to the DNA, whose entire sequence is called the "genome." Portions of the genome are active, others not. If the gene attaches to an inactive portion, the host's antibiotics kill it. If it attaches to the active portion, it may be positioned such that a defective organism is created, and one, as well, that may cause disease. With microinjection—the most common method of creating transgenic animals—only one to two percent of injected mammalian eggs result in a birth. And of those few that are born, only a small percentage have the transgenic gene integrated into its DNA. But most of those are prone to disease. What more do we need to tell us that the project runs counter to the way of Nature and is therefore radically perverse?

In addition to the presence of vectors, other problems come from the presence of other genes, including viruses, which are usually inserted with the transfer gene. One class of these genes are markers, which confer resistance to antibiotics. Because by itself there would be no way to tell if the transfer gene had been successfully inserted, markers are attached to it. Once the insertion has been made, the recipient cells are tested with an antibiotic. Those cells which have integrated the antibiotic resistant gene onto its DNA will survive, and scientists therefore assume that the desired gene most likely has been integrated too.

Another class of attached genes are promoters, which are added to ensure that the desired gene expresses itself consistently and strongly. Promoters are present in all DNA, and in naturally occurring situations they are under the control of regulator genes. But regulators are not inserted with promoters and the desired gene.

Finally, there is a class of barrier penetration genes, whose function is to overcome the natural resistance that cells have to foreign genes attaching to its DNA. A cell's enzymes will attack an alien gene and cut it up or otherwise disable it. A combination of genes are used by scientists to overcome this barrier.

There are serious problems that come with using viruses and bacteria for vectors, and with using genes in the insertion package. Mae-Wan Ho, a British microbiologist who has written widely on this subject, has isolated eleven negative consequences of genetic engineering. The following impact human health directly.

1) "Horizontal gene transfer" refers to the transfer of genes across species barriers. Because the vectors used for

insertion are selected for their ability to transgress species boundaries, they can move from the host species to others. Dr. Ho notes that " . . . [of] 75 references published in mainstream journals between 1993 and 1996, all but 2 [give] direct or indirect evidence of horizontal gene transfers." According to Ho, "a gene transferred to any species in a vector can reach every other species on earth, the microbial/viral pool providing the main genetic thoroughfare and reservoir." As an example, she notes that a mobile genetic element first found in fruit flies was discovered in humans. Scientists speculate that the fly's gene entered a virus which effected the transfer.

2) Using plasmids as vectors will spread antibiotic resistance, already a major health crisis. Transgenic tomatoes, for instance, carry genes that are resistant to kanamycin, the bacteria used to treat tuberculosis. With increasing cases of tuberculosis being recorded, this is not good news. Other antibiotic resistant genes will nullify other antibiotics.

3) Vectors, she says, may cross species boundaries and recombine "to create new pathogenic bacteria and viruses." The disabled viral vectors will also "recombine with a wide range of natural pathogens." Even though 'crippled,' these viruses can combine with others and with mobile genetic elements "to jump in and out of genomes. Otherwise it would have been impossible to construct any transgenic organisms at all."

4) Eating transgenic food may allow the vector to integrate itself into the host's genome. The host organism may then provide an environment in which the virus can reverse its disabled pathogenic properties.

5) Some transgenic plants, such as the soybean with a brazil-nut gene, are engineered with allergens, some of which can be fatal.

6) Once the vectors are in the environment, they cannot be controlled and may persist indefinitely.

To give just one example of the virulence of viral vectors, consider that the one used to transmit genes to fish is composed of three viruses: one of which is known to cause sarcomas in chickens; another of which causes oral lesions in humans, as well as pigs, cattle, and horses; and the third of which creates leukemia in mice.

Even the scientists on the Food and Drug Administration's Microbiology Group warned the agency that it was making a mistake by " . . . trying to force an ultimate conclusion that there is no difference between foods modified by genetic engineering and foods modified by traditional breeding practices." Further, it wrote, " . . . they lead to different risks." One group member noted: "There is a profound difference between the types of unexpected effects from traditional breeding and genetic engineering which is just glanced over in this document."

Nobel laureate and biologist George Wald warned against the perils of genetic engineering many years ago. In a 1967 essay he wrote: "Up to now, living organisms have evolved very slowly, and new forms have had plenty of time to settle in. Now whole proteins will be transposed overnight into wholly new associations, with consequences no one can foretell, either for the host organism, or their neighbors . . . . going ahead in this direction may be not only unwise, but dangerous. Potentially, it could breed new animal and plant diseases, new sources of cancer, novel epidemics."

5.

There are three reasons for genetically engineering species:

1) To "harvest" a by-product from the new organism. For example, when the human insulin gene is inserted into an e.coli bacterium, the bacterium, in reproducing, creates more insulin.

2) To protect an organism against pests or the environment. An antifreeze protein from flounder has been injected into Atlantic salmon to allow them to live in colder waters.

3) "To improve the species." A growth hormone from rainbow trout, for example, is inserted into carp or catfish embryos to speed their growth.

Susan Aldridge characterizes novel species as " . . .

somewhat different from the creatures that have emerged during the course of evolution. How radical this difference is depends on your viewpoint." Absolutely. And since we, as a society, have no set of common principles or values, the law has no built-in assumptions by which to oppose this technology. Instead, when the first genetically engineered form of life was patented in 1980, the United States Supreme Court upheld the legality of the action. The mind boggles at the presumption of someone who would design a new species, and at that of the court which supported the "right" to own and earn a profit from it, especially as it opens the prospect of human genetic engineering.

6.

Philosophically the most interesting insight from genetic engineering is also an argument against its practice. It turns out that genes are not, as microbiologists first thought, simple constants in a field with other constants. That is, a gene does not express itself in isolation from other genes; we cannot expect a particular gene, by itself, to express a particular trait. Rather, what the gene expresses is affected by other genes in the organism and even the environment. The reductionist model does not work for genes.

Thinking of the gene as a stable, unchanging building block that will invariably express itself the same way is an instance of the rationalist vision of the world as a collection of boxes. In this case the boxes are genes and organisms: the genes are the building blocks (little boxes) of all organisms (bigger boxes) which can be taken apart and put back together in novel combinations with clear predictability. However, nature does not work that way. Jaan Suurkula, M.D., drawing on Mae-Wan Ho's book, *Genetic Engineering: Dreams or Nightmares?*, notes that genes transferred to plants have created unpredicted allergens and toxins, and that certain transgenic plants, including cotton, tomatoes, and corn, have not grown well or grown at all in regions or countries far from where they were developed. Finally, genetically built-in Bt-toxin, used to guard plants from insects, was almost virtually useless by the second generation of insects introduced to it: 70 percent had developed Bt resistance. Clearly the parts of the world are not as discrete as we thought.

In *Genetics & The Manipulation of Life: The Forgotten Factor of Context*, Craig Holdrege makes the same point as Mai-Wan Ho, with examples that readily demonstrate the importance of environment on plant growth and form. In one instance, the same species of dandelion grown in a yard, a garden, and a forest (all within seventy meters of each other) exhibited different traits. The yard dandelion had leaves growing horizontally, with its flower head no higher than the grass; the garden dandelion had nearly upright leaves and taller, thinner stalks; those in the woods had leaves and flower heads much larger than either of the other two.

For his next instance, Holdrege cites the experiment of French scientist Bonnier who took young plants of many different species, cut them in half, and planted half of each in a garden at 6,562 feet above sea level and the other half in garden soil near Paris. Holdrege notes that "every part of the plant—down to the details of cellular structure—changed." The alpine plants had strong root systems, small leaves, and one or two flower heads, while the plants near sea level had fuller leaf and flower development and a relatively less developed root system.

Holdrege's third instance is an experiment in which soil is the variable. He grew three species of plants—groundsel, rape, and broad bean—in garden soil and in an equal mixture of loam and sand. Each of the three species grown in one soil developed similar characteristics, markedly different throughout their growth and maturation, from their counterparts grown in the other. The mature broad bean, groundsel, and rape grown in sandy loam all had long secondary roots, with small tertiary branches. By contrast each of the three species grown in composted garden soil, when mature, had extensive, complex root branching. And while the plants grown in garden soil had multiple stems and abundant leaves, the plants grown in sandy loam had only one stem and smaller leaves.

Holdrege rightly concluded that the "environment draws forth the same tendency in each plant" and that every part of the plants reflected their "environmental condition as a whole." The plant is not a mere machine programmed from a cotyledon to a predetermined form regardless of the quality of the soil or light or density of air. Each of these factors is, in his terms, relational. The plant that develops small leaves and flowers on a mountain side is telling us about the effect that altitude has on it. And so does every other environmental factor that impacts a plant's growth and form. We learn from Holdredge's observations not only about the plant but about the variable: that how a plant grows and the forms it takes depend upon its environmental context.

In the late eighteenth century, Goethe made similar observations in the Alps. "Whereas in the lower regions, the branches and stalks were stronger and more massive, the buds closer together, and the leaves broad, higher in the mountains the branches and stems were more delicate, the buds farther apart, so that a greater space occurred from joint to joint and the leaves took on a more spear-like shape. I noticed this in the case of a willow and a gentian, and was convinced that they were not different kinds. Also, by Walchensee I noticed longer and slenderer rushes than in the lowlands."

What all these examples illustrate, as Holdredge points out, is that plants possess plasticity, a play between species specificity and individuation. If genes were an absolute determinant, variations within varieties, such as the above examples, would be impossible. But given what bioengineers wish to accomplish, it is clear that even today genetic theory derives from the methods of Gregor Mendel, founder of genetic thory, as recorded in his classic paper, "Experiments on Plant Hybrids." Based on eight years of work with hybrid peas and published in 1866 when Mendel was abbot of a monastery in Moravia, the paper opens with a statement of his intent "to determine the number of different forms in which hybrid progeny appear . . . [classify] these forms in each generation with certainty, and ascertain their numerical interrelationships." Mendel wanted certainty, and certainty for Mendel, a trained physicist, came through the application of number to natural phenomena.

To achieve this certainty, however, Mendel had to limit his observations. In the first place, he had to choose plants which, when self-pollinated, would repeatedly yield distinct traits. ("The experimental plants must necessarily . . . . possess differing traits.") Plant breeders knew that when two plants with common traits were crossed that their progeny and subsequent generations would retain those traits. If, in addition, these plants had differing traits, their hybrids and subsequent generations would form a new trait. Mendel's experiment was designed "to observe these changes for each pair of differing traits, and to deduce the law according to which they appear in successive generations."

Mendel chose from among those pea plants with "differences in length and color of stem; in size and shape of leaves; in position, color, and size of flowers; in length of flower stalks; in color, shape, and size of pods; in shape and size of seeds; and in coloration of seed coat and albumen." The next two sentences betray the weakness of the project. "However, some of the traits listed do not permit a definite and sharp separation, since the difference rests on a "more or less" which is often difficult to define. Such traits were not usable for individual experiments; these had to be limited to characteristics which stand out clearly and decisively in the plants." "More or less" would not do because a trait that is "more or less" brown or round cannot be categorized and therefore cannot be quantified. To further ensure certainty, Mendel decided on a third limiting factor: to include only vigorous plants in his experiments. "Weak plants always give uncertain results . . ." Their offspring either fail to flower or, if they do, produce only a few inferior seeds. In these ways, then, Mendel has eliminated phenomena that would have made it impossible for him to express his findings in a ratio of approximate whole numbers.

Mendel began the experiment by observing the following traits: 1) shape of ripe seeds; 2) color of seed albumen;

3) color of seed coat; 4) shape of ripe pod; 5) color of unripe pod; 6) position of flowers; and 7) stem length. In some cases the differences between traits were clearly marked. For example, ripe seeds were either round or nearly so, or else angular. But not all traits exhibited that approximate uniformity. For example, the color of the seed coat was either white; or it was gray, gray-brown, or brown "with or without violet spotting." What Mendel discovered was that the hybrid trait was not an intermediate between the two parental traits, but one or the other (or so close as not to make a difference). The traits that appeared in the hybrid he called the "dominating" traits. (They are now termed "dominant.") When the hybrid plants were self-pollinated, both traits from the parents of the hybrids reappeared in the progeny, in ratios corresponding very closely to 3:1. In the second hybrid generation, those plants whose parents exhibited the recessive trait also had it. Of those with parents exhibiting the dominant characteristic, these had dominant to recessive traits in a ratio of 3:1.

Mendel's focus on isolated characteristics is typical of the technical mind that seeks to classify parts qua parts, rather than perceive parts in relation to wholes. The distinction is significant, and it has radical consequences for us. It is founded upon an insufficient understanding of what constitutes a whole. In mathematical terms a whole is equal to the sum of its parts, but in organic terms parts of a living system considered in relation to each other create something beyond a mere aggregation.

The whole of a plant, Holdrege reminds us, exists in time as well as in space: one cannot isolate its form at any stage of its existence and identify it with the plant in its totality. The process of growth, maturation, and decay, with accompanying changes in leaves, stems, and roots remind us that the individual plant or animal is a process, not a static entity. Furthermore, each plant, each creature, is endowed with limits of form and other traits, but that these are not absolute determinants. As we saw with Holdrege's dandelions and Bonnier's plants, environments can evoke wide variations within species, but variations exist also within time, from a plant's germination to its death. Heredity and environment are each partial determinants of growth and form. But as Holdrege emphasizes, the geneticist is not concerned with the organism as a whole but with rigid characteristics that can be quantified, thus reducing the organism to a mere fraction of what it truly is. This is precisely what technique does: reduce an organism or natural or social process to its simplest operations and obvious characteristics, making it more susceptible to manipulation and control.

8.

Because of its control over public information, perhaps no technique has as much influence as journalism. The journalist must deliver his story quickly and it must be immediately comprehensible; for this reason he must simplify. Like anyone else delivering information, he can distort the facts by eliminating those with which he or his employer do not agree.

"Harvest of Fear," a television documentary on genetically engineered foods produced by *Frontline*, dealt with public reaction to genetically modified crops. Its first half was an even-handed presentation, balancing public fears and reasoned opposition to genetically modified crops with biotech industry arguments for them. But an interview with Norman Borlaug, Nobel Prize winning agronomist at Texas A&M University, whose tireless work has helped reverse crop shortages in the Third World, introduced a change in tone. "This organic movement," Borlaug announced, "is ridiculous." The documentary then presented three case studies to demonstrate that biotechnology is necessary to save crops, and not just in the Third World.

The first example focused on the papaya, the basis of a $45 million Hawaiian crop which was being attacked by a ring spot virus poised to destroy the entire industry. According to the documentary, neither physical barriers nor chemical pesticides had been of any avail. Cornell University researcher Dennis Gonsalves found that by splicing a gene from the ring spot virus—a gene "that made a harmless protein in the virus's outer coat"—into the papaya DNA, that he could protect the fruit against infection.

In the second case, a Kenyan technician, Florence Wambugu, looked for a way to improve the sweet potato crop in her home country. Like the papaya, the sweet potato was attacked by a virus, which in this case created a smaller, paler sweet potato. Here, the documentary said, was an example of the inability of organic agriculture to produce a crop capable of feeding a growing population. Current African farming relies upon traditional agricultural methods of production, which means it does without chemical fertilizers or pesticides. While these methods saved the crop, according to Wambugu, they did nothing to create a particularly healthy vegetable or to prevent the spread of the feathery mottle virus. Wambugu appealed to Monsanto, and at their lab she developed a bioengineered sweet potato with healthy size and color.

In the third case, Luis Herrera-Estrella, a researcher at the Biotechnology Center in Irapuato, Mexico, tackled the problem of acid soil and aluminum toxicity, which was reducing crop yields. Herrera-Estrella has dedicated himself to creating bioengineered plants that are tolerant to aluminum toxicity. The problem was that soil acidity near Irapuato was creating aluminum toxicity at the same time as it was keeping phosphate, a necessary nutrient, in insoluble form. Since aluminum tolerant plants secrete higher amounts of citric acids than less sensitive varieties, Herrera-Estrella inserted a gene for citric acid production into tobacco, rice, and maize. The citric acid detoxified the aluminum at the same time as it made the phosphate soluble. According to his records, his genetically engineered tobacco planted in Irapuato soil, with little fertilizer, grew larger leaves than conventional varieties grown in the same soil. His maize was more successful too.

In the case of Hawaiian papayas and Kenyan sweet potatoes we have monocultures attacked by a virus. When one thinks about the totality of a plant's environment—soil; the presence or absence of bacteria, fungi, and earthworms; light rhythm; surrounding plants; rainfall frequency; water quality—one begins wondering if altering or removing one or more of the variables for long periods might be sufficient to account for a virus's sudden appearance and take over. And what caused the acid soil? The *Frontline* documentary did not address these problems; it simply took the word Monsanto and Cornell University spokesmen that organic agriculture was incapable of addressing these issues.

The answer to a virus epidemic comes from China, where scientists under the direction of Y. Zhu worked to find a natural solution to a rice crop attacked by a fungus. The solution the Chinese scientists discovered was to grow several varieties of rice rather than one. As agronomist Martin S. Wolfe noted in his report of Zhu's work for Nature, "Until about 100 years ago, monoculture was practiced only at the level of species, with, for example, wheat, maize or rice becoming dominant in different climatic regions. Monoculture has since expanded to different levels, reducing the number of species, of varieties within species, and particularly of genetic differences within varieties." The problem with the current monoculture is that one variety grown exclusively can be destroyed (as experience with the Hawaiian papaya demonstrated) by one pathogen. Or, to paraphrase Wolfe, a pathogen that destroys one plant within one variety of a monoculture has the potential to destroy the entire monoculture. Agribusiness's response is to develop varieties resistant to the pathogen or a pesticide that will kill it. But as experience with Bt corn has shown, resistance is short lived. By no more than two generations, a new strain of pathogen develops that infects the monoculture.

The Chinese experiment demonstrated that the several varieties grown acted for each other as a physical barrier against the fungus. It also appears that when a plant is attacked by a pathogen to which it is immune, the attack stimulates its immune system, which then makes it better able to withstand an attack by a pathogen to which it is susceptible. In addition, Zhu and his colleagues speculate that by growing several varieties of rice in a single field, the pathogens were forced to compete with one another, thus slowing each one's adaptation to the different varieties. The possibility then exists that "different mixtures of varieties in different fields in different years could slow down the adaptation of the pathogen even more." Zhu's experiment demonstrated that by increasing the mixture of varieties, fewer epidemics swept the fields. As a consequence, the farmers stopped applying fungicides.

Wolfe's own experiments with Wolfe's own experiments with mixing wheat varieties have demonstrated that diseases attacking any one of the varieties restricts the spread of disease throughout the mixture. Wolfe concluded that "Variety mixtures may not provide all the answers to the problems of controlling diseases and producing stable yields in modern agriculture. But their performance so far in experimental situations merits their wider uptake."

As for "Harvest of Fear"'s third case, the acidic soil near Irapuato, Mexico, that turns out to be a man-made problem. In fact, approximately 40 percent of the world's arable land has such a high acid content that some say it poses a serious problem for crop yields. Acid soil comes from acid rain and from so-called conventional agricultural practice, which relies upon high amounts of chemical fertilizers. These contain acids that lower soil pH. Usually lime, dolomite, and gypsum are used to neutralize the pH or to increase alkalinity. In Australia, volcanic dust is being used to the same effect.

But bioengineers and others creating an agriculture of scale choose not to follow the way of nature. They do not want to work with nature, as they claim, but seek to control nature by short cutting certain processes, as the techniques for gene insertion demonstrate. Instead of recognizing the diversity of plants, animals, bacteria and virus as necessary parts of a interdependent whole, they seek to transform parts into independent wholes. As Holdrege and co-author Steve Talbot write in "Sowing Technology," agronomists and others creating our agriculture of scale, seek to reduce each monoculture to a uniform variety. Speaking of the vast Nebraska cornfields they write: " . . . it is difficult to find much of nature in those cornfields. . . . today's advanced crop production uproots the plant from anything like a natural, ecological setting. This, in fact, is the whole intention. Agricultural technology delivers, along with the seed, an entire artificial production environment designed to render the crop independent of local conditions."

9.

Since technique seeks to dominate every aspect of the economy and culture, it is not surprising that industrialization is spreading to the organic food sector. Some organic food companies are adopting agribusiness's processing techniques. Horizon, for example, which controls 70 percent of the organic milk market, uses ultrapasteurization, which eliminates vitamin and enzymes but preserves the milk, allowing it to be shipped long distances. And Cascadian Farm, one of the country's largest organic food companies, produces processed organic TV dinners with additives.

A decade ago, when sales of organic foods accounted for less than 1 percent of the country's total food sales, agribusiness did not take the organic movement seriously. Now, with sales growing by 20 percent every year for the last decade, the market is attractive. General Mills, for example, acquired Cascadian Farm. Cascadian Farm was started in 1971 by former back-to-the-lander Gene Kahn, who began growing organic food for a hippie collective near Bellingham, Washington. Michael Pollan has traced Kahn's story from the founding of Cascadian Farm to Kahn's current position as a vice president of General Mills. Fairly early in the business Kahn discovered he could increase sales and profits if he contracted others to grow the food he sold. Eventually Cascadian Farm industrialized much of its operations and even developed organic TV dinners. For many, a processed organic TV dinner is a contradiction in terms, but it illustrates what is happening to the organic movement.

Fred Kirchenmann, director of the Aldo Leopold Center for Sustainable Agriculture, notes that the organic movement originated in reaction to industrial agriculture. "Those folks who said, 'This is the wrong thing to do,' are the ones who started the organic movement. They invoked the law of return. The law of return says that if you take something from the land, you have to bring something back.

"Because it has its origins in this moral mandate, organic as a movement, from its roots, had a moral obligation. A lot of the early folks in the organic movement were motivated by that sense. It was almost missionary."

As the movement grew, the organic industry emerged. Kirschenmann notes that conflict naturally developed between the moral mandate and the need for sales: "Whether or not these two motivations are compatible is an unanswered question, but it's fair to say that the majority of people in the movement think there's a need for the industry." Still, Kirchenmann says, they don't want to lose those things that originally attracted them to organic agriculture: its land stewardship, its support of the family farm, and the quality of the food it produces.

As the organic movement grew, the need for standards appeared as increasing numbers of people bought organic meat and produce from faraway growers and needed assurance that what they were buying was in fact organic. Out of the need for assurance came organic certification, which calls for certifying entities. With a number of certifying entities across the country and variations in standards, the USDA in 1991 created the National Organic Standards Board. In 1997 the board issued a set of proposed standards which former board member Bill Welsh says were fine, but were "so full of loopholes that they became valueless." The rules made it clear, for example, that antibiotics should not be used in organic agriculture but then went on to cite instances when it was acceptable. Another loophole allowed sludge to be used for fertilizer.

It was not agribusiness but the Organic Standards Board itself that was responsible for what Welsh calls "a farce, a complete farce." Board members wanted exceptions that would suit their own needs. One of them, said Welsh, didn't care what was in the meat he sold "as long as it had a USDA label and he could sell it to Japan."

Fortunately, the rules were subject to ninety days of public comment, and comment the public did. Loudly. So vociferous and widespread was public opposition that the board withdrew its proposed rules and went to work again, eliminating the most egregious loopholes. Gene Kahn, who was on the board between 1992 and 1997, lobbied successfully for additives and synthetic chemicals, which remained in the final standards. "If we'd lost on synthetics," Kahn told Pollan, "we'd be out of business."

These exceptions, as Pollan wrote, favor the large organic companies. Considering as well the fact that these companies are buying from industrial organic farms and not from small family farms, it is not surprising that some organic growers are refusing the organic label. It is once again a case of big absorbing small, and of the triumph of technique.

CONCLUSION

To make its point this book could have been written about almost any other manifestation of technique, for all end at the same point: the elimination of human impulse in favor of the machine and uniformity. But the focus on contemporary agriculture has the advantage of leading us to reflect on the material bases of life—our dependence on finite resources, the condition of those resources, their relation to the quality of our food, and the relation of food to health. As we have seen, driven by technique, industrialized agriculture relies upon constant development of new products and technologies, as does the whole of industrial capitalism. As mentioned in the introduction, innovation and new product development are necessary for any company wishing to remain competitive and to continue generating profits. But technique, as we also saw, generated the centralized economy, using direct mail marketing, railroads, and mass production as its means.

Today all facets of the Western economy and culture are centralized to the degree that every marketable product from clothes to houses is standardized. We pride ourselves at being on the apex of human development, that our bourgeois condition is that to which humanity has striven for millions of years. So little do most Americans know of history and past cultures that they may well believe that our degree of centralization represents a healthy state of affairs.

Look at the extent to which the comforts of the modern American depend upon centralization: his electricity is generated by a plant that supplies power to thousands, usually hundreds of thousands and even millions of users; his long-distance telephone service is most likely supplied by one of three communications giants; his television, his main instrument of "relaxation," is manufactured by one of a dozen companies; his favorite magazine is one of perhaps a score published by a corporate giant. His car is manufactured by one of a dozen companies and his gasoline is processed by one of another handful. When he goes to see a movie, it may be one that has been made and distributed by the conglomerate that publishes his favorite magazine. In any facet of the economy, oligopoly reigns. As the computer integrates more and more activities and instruments, and conglomerates continue horizontal and vertical integration, centralization is intensified.

At first sight, decentralization appears to be the answer to technique and the concomitant dehumanization of society. But decentralization will only thrive when technique is under the control of man and no longer acts as an automatism. A decentralized economy in which man controlled technique would be a no-growth economy with many centers of power production, manufacture, food processing, and so on. But the political history of the West since the fourteenth century has been just the reverse: smaller units aggrandizing themselves into larger ones: cities uniting into regions, regions into nation states, and these sending emissaries and armies overseas in quest of empire. While states created by colonial administration and communist dictatorships are devolving back into tribal and ethnic units, the imperial imperative is alive in the West. The West's empire today is economic, backed by military might. The root of its empire lies in efficiency and large scale production, a vast enterprise uniting millions of managers and technicians in all fields, harnessing science and scientism for the sake of profit and power.

2.

Dostoevsky wrote that without God all things are possible. For at least a hundred years Western culture has been in the process of demonstrating the truth of that claim. As the practical applications of science created material explanations for more and more dimensions of reality, people were induced to believe that science and man's intellect were sufficient for the understanding of all things. For the man-in-the-street the mystery and fear of the transcendent diminished once his physical comforts and civilization's power over Nature increased. Eventually, philosophical materialism and scientism, along with hatred of religion, was embraced by many, particularly by intellectuals for whom these became pillars of a counter-religion. In consequence, the West created an almost entirely profane culture.

There we have the underlying reason why we cannot throw off technique: in the absence of the sacred the West necessarily had to embrace a substitute. The Myth of Progress, with its doctrine of the continual betterment of humanity through technical achievement, is precisely that. For that reason the simplification of our system is out of the question. Individuals here and there may choose to opt out of the system, but the system will never reject technique and re-organize itself. In the first place, there is too much money invested in the present order and the individuals and corporations with money are not about to relinquish it; second, too many people are addicted to comfort and have become incapable of knowing their condition of servitude; third, the Myth of Progress is far from dead, and tens of millions still believe that technical advances will some day usher in the Golden Age.

Lewis Mumford once expressed to me that only the appearance of another axial figure would rectify affairs. By axial figure he meant a revolutionary individual of the size of Jesus, Lao Tze, the Buddha, or Confucius. I would add that only another infusion of the Spirit through such a figure would renew this world. A radical renewal would crack the bonds of materialism that bind our minds like iron cords, growing like tendrils over our institutions and books. With William Blake's "forg'd manacles," we have chained the Imagination. We are no longer able to see things as aspects of Infinity, but see all things as finite and measurable. Concomitant with the loss of intellectual intuition or Imagination is the separation from Eden: the loss of the Infinite, a forgetfulness of the Center. To compensate we sought power in this world, a control over Nature. We divided the world into parts: intellectuals devised more and more categories to mangle experience and further alienate us from that which Is. In the process more and more chains were forged.

In truth, the entire thrust of this "civilization" is aimed at the overthrow of meaningfulness. The triumph of technique is precisely that. Whether it be the bioengineering of human beings, plants, and animals or the extermination of thousands of enemy troops, the quest is for infinite power. It seems a certainty, however, that given its ever-increasing technical achievements, our civilization cannot long endure. Each new major breakthrough ensures greater anxiety and uncertainty as we create an increasingly artificial world controlled by rational processes. The more complex and artificial the individual subsystems become, the more the entire system is destabilized and liable to break down. This book examined the subsystem known as agribusiness and illustrated how increasing yield through pesticides and chemical fertilizers has degraded water and soil, and caused top soil loss. Those pesticides using dioxins are causing reproductive failures in animals and very possibly in humans too. We looked at bioengineering and the potential threats it poses, both to other species and humanity itself. We saw that creating a uniform plant for monocultures degrades the soil, making it susceptible to outbreaks of epidemics by pathogens. Our ability to use science successfully within limits has made us arrogant. We have reached the point at which the society as a whole is incapable of taking self-corrective measures.

3.

Many years ago English novelist and playwright J.B. Priestley wrote a wise epilogue to his survey of Western literature, *Literature and Western Man*. There Priestley pointed out that in the Renaissance, Western Man's psyche was balanced between his conscious and unconscious, a healthy condition in which conscious control did not overbalance the impulses, symbols, and demands of the unconscious. "After the loosening and decay of the medieval religious foundation and framework," Priestley wrote, "Western Man broke out, shook himself free. . . . His outer and inner worlds were not yet at variance; the religious symbols had not yet lost their force. The greatest writers of this age . . . are nicely balanced between the conscious and unconscious life. . . . These writers seem neither over-extroverted nor obviously introverted. Their age is neither committed to one conscious attitude nor at the mercy of its unconscious drives and urges . . ."

One-sidedness overtook the West with the Enlightenment, when mathematics and science provided us with instruments to redirect and thus distort natural processes. Through science the Myth of Progress was born, and in that myth science became the most efficacious means for achieving happiness. Scientism, the belief that all fields of inquiry should be subordinated to science and mathematics and interpreted in their light, was also born. The control of all activity by reason seems eminently sensible, but it was achieved through the suppression of the unconscious. But the unconscious never loses force and influence; if suppressed, it reappears in destructive impulses. The Enlightenment project was born of the urge for measure and order, a desire not only to reveal a machine-like regularity in the operations of Nature, but to extend these presumed principles of Nature to every facet of human affairs. But machine-like order and regularity in human activity and institutions can only be guaranteed by total centralized control. During the Enlightenment, society's religious framework began disintegrating, as did its traditional class structure. In the following century, the Industrial Revolution created the urban mass. As Priestley wrote: "Patterns of living that had existed for thousands of years are destroyed within a generation. . . . Meanwhile, the inner world is largely ignored, the unconscious drives and fantasies remain unchecked."

Priestley's argument is repeated and amplified in *The Pentagon of Power*, Lewis Mumford's last great work. In a section titled "Addled Subjectivity," Mumford argued that the development of language, ritual, and custom were responsible for bringing primitive man's unconscious under control. It is man's gift for "exact repetition," Mumford wrote, " . . . that lies at the bottom of human culture. . ." Repetition has enabled man to build an internal structure of meaning and routine. It is this routine, which Priestley also cites, that curbed the promptings of the unconscious. But in our time, wrote Mumford, "the Power Complex has not merely deliberately disrupted salutary customs and undermined traditional moral values: what is even more serious, it has transferred all the stabilizing repetitive processes from the organism to the machine, leaving man himself more exposed than ever to his own disordered subjectivity. . . . As a result, the unconscious has now resumed its early dominion over man." As Mumford goes on to write, humanity's present situation is greatly exacerbated by the fact that it now has enormously powerful weapons of destruction at its command.

Many of the establishment's fantasies seek to develop large scale projects, which is a mark of empire. The mantra, Bigger is Better, still holds true in the minds of the technocrats and managers running the system. For example, the federal government spends massive amounts of money on a program that will enable it "to conquer space," while we haven't the slightest idea how to deal with our inner demons; it leads us to propose a Star Wars missile defense system, even in the wake of the attacks against the World Trade Center and the Pentagon, in which the industrial-military complex was stunned and hammered by three airliners and less than two dozen hijackers. Irrationality is evidenced in the entire "conventional" system of agriculture erected after World War II: in the heavy use of chemical poisons and fertilizers, in the drive to create monocultures of one variety that can be grown anywhere irrespective of environment; in the endeavor to bioengineer plants and animals in disregard of natural law or cautionary safety; in the rush to clone animals solely to harvest their organs for medicine: each of these is symptomatic of a society in which any scientist's bizarre fantasy can become life, provided it is technically feasible and stands the chance of turning a profit. As though it were devised solely as an illustration for this book, Iowa farmer Tom Dorr, President's Bush's nominee for Undersecretary of Agriculture for Rural Development, proposes that farms be enlarged to an average of 225,000 acres, or one for every 350 square miles. These, of course, would be corporate owned and centrally managed. Such a vision would reduce farmers to hired hands, destroy rural towns and villages, and put the entire food supply in the hands of relatively few corporations.

A system in which avarice and the lust for power are the grease that keeps things running is a system that will collapse, sooner rather than later. The argument for the collapse is at least as old as Plato, and it is found in the first book of *The Republic.* There Socrates argues with a sophist named Thrasymachus, who maintains that justice is the interest of the stronger. In the course of the argument Socrates gets Thrasymachus to admit that the unjust man demands more than his due, both of the just and the unjust. The obvious conclusion that Socrates draws is that a city or army or any group of unjust men will not be able to act in concert, for each will try to outdo or get the better of the others. He concludes: "For factions, Thrasymachus, are the outcome of injustice, and hatreds and internecine conflicts, but justice brings oneness of mind and love. . . . If it is the business of injustice to engender hatred wherever it is found, will it not, when it springs up either among free men or slaves, cause them to hate and be at strife with one another, and make them incapable of effective action in common?" Doesn't this mirror our situation exactly?

For Plato justice is the due measure between elements: within the individual it is the rule of reason over will and appetite; within the state it is the rule of the guardians over the soldiers, artisans, and farmers. It is the rule of wisdom and goodness, and to break the hierarchy is to engender chaos. But for the Power Complex the rule of the wise and good is a joke, for the managers and technocrats are on the side of Thrasymachus, and like the Athenian sophist they believe that the just man is a simpleton; unlike Thrasymachus, however, they come armed with promises of Heaven on Earth, while what they create is at the opposite extreme.

Their victory, which is inevitable, can be but transitory, and in the aftermath of this civilization's collapse we will be glad of the decentralist experiments that are being carried out across the nation: in local food systems, organic agriculture, alternative energies, barter systems—in all the projects that restore human scale to human enterprise, that emphasize cooperation over competition, virtue over efficiency and control. Regionalism, which is a form of decentralization, is our only counter to globalization, and while it is not sufficient by itself to reestablish a humane society, it is certainly a necessary precondition. It is also an alternative to the homogenization of the American landscape and culture, and of all societies worldwide. Vincit omnia veritas.

BIBLIOGRAPHY

Aldridge, Susan. *The Thread of Life: The Story of Genes and Genetic Engineering*. Cambridge: Cambridge University Press, 1998.

Alliance for Bio-Integrity. "FDA Documents Show They Ignored GMO Safety Warnings from their Own Scientists." online.sfsu.edu/-rone/GE%20Essays/FDAdocuments.html.

__________. "Mother Jones Releases USDA Memo Detail - ing Plans to Gut NOSB Recommendations on Organic Standards." www.monsantosucks.comusdaLeak.html. 1998.

Antoniou, Dr. Michael. "Genetic Engineering and Traditional Breeding Methods: A Technical Perspective" (excerpt). www.psrast.org/mianunpr.htm.

Beard, Charles and Mary. *The Rise of American Civilization*. New York: The MacMillan Company, 1930.

CBS News. "60 Minutes": Mike Wallace story on the American Farm Bureau. Transcript. Aired April 9, 2000.

Coburn, Theo, Dianne Dumanoski, and John Peterson. *Our Stolen Future: Are We Threatening Our Fertility, Intelligence and Survival A Scientific Detective Story*. New York: Dutton, 1996.

Committee on Economic Development. "An Adaptive Program for Agriculture." 1964.

Corcos, Alain F. and Floyd V. Monaghan, *Gregor Mendel's Experiments on Plant Hybrids: A Guided Study.* New Brunswick: Rutgers University Press, 1993.

Defenders of Wildlife. "Amber Waves of Gain: How the Farm Bureau is Reaping Profits at the Expense of America's Family Farmers, Taxpayers and the Environment."

Department of Economics, Iowa State University. "Census of Agriculture." www.profiles. iastate.edu/data/census /county/agcensus.asp?sCounty=19000.

Eliade, Marcea. *The Sacred and Profane*. New York: Harcourt, Brace & World, Inc., 1959.

Ellul, Jacques, *The Technological Society*. New York: Vintage Books, 1964.

Epstein, Ron. "Redesigning the World: Ethical Questions About Genetic Engineering."online.sfsu.edu-rone/GE%20Essays/Redesigning.htm.

Goethe, Johann Wolfgang. *The Metamorphosis of Plants*. Kimberton, PA: Bio-Dynamic Farming and Gardening Association, 1993.

Hall, Sam B. "The Truth about the Farm Bureau." Denver: Golden Bell Press, 1956.

Ho, Mae-Wan. "Horizontal Gene Transfer—New Evidence." www.greens.org/s-r/24/24-25.hml.

_________. "The Unholy Alliance." *The Ecologist*, vol. 27, no. 4, (July/Aug, 1997).

_________ and Beatrix Tappeser. "Transgenic Transgression of Species Integrity and Species Boundaries." userwww.sfsu.edu.

Holdredge, Craig. *Genetics and the Manipulation of Life: The Forgotten Factor of Context*. Hudson, NY: Lindisfarne Press, 1996.

_____________ and Steve Talbott. "Sowing Technology: Do We Really Want to Pit Technology Against Nature?" Ghent, NY: The Nature Institute, 2001.

Lauck, Jon. *American Agriculture and the Problem of Monopoly*. Lincoln: University of Nebraska Press, 1999.

________. "Toward an Agrarian Antitrust: A New Direction for Agricultural Law." *North Dakota Law Review* 75, no. 3. (1999).

LeGoff, Jacques. *Your Money or Your Life*. New York: Zone Books, 1988.

McMath, Robert C. Jr. *Populist Vanguard: A History of the Southern Farmers' Alliance*. Chapel Hill: University of North Carolina Press, 1975.

Mendel, Gregor. "Plant Hybridization." Harvard University Press, Cambridge.

Mumford, Lewis. *The Pentagon of Power*. New York: Harcourt Brace Jovanovich,1970.

National Post Online. "The trouble is, this cure might kill you." www.nationalpost.com/content/features/genome/0314005.html

Palfreman, Jon. "Harvest of Fear." Script for Frontline/Nova special presentation. Original air date 4/23/01.

Physicians and Scientists for Responsible Application of Science and Technology. "What Is Genetic Engineering? A Simple Introduction." www.psrast.org/whatisge.htm.

___________. "The New Understanding of Genes." www.psrast.org/ newgen.htm.

___________. "How Are Genes Engineered?"www.psrast.org/whisge.htm.

Pollack, Robert. *Signs of Life: The Language and Meanings of DNA*. New York: Houghton Mifflin Company, 1994.

Priestley, J.B. *Literature and Western Man*. New York: Harper & Brothers, 1960.

Rifkin, Jeremy. *The Biotech Century: Harnessing the Gene and Remaking the World*. New York : Tarcher/Putnam, 1999.

Saul, John Ransom. *Voltaire's Bastards: The Dictatorship of Reason in the West*. New York: Vintage Books, 1993.

Suzuki, David and Peter Knudtson. *Genethics: The Ethics of Engineering Life*. Toronto: Stoddart, 1988.

Schwab, Jim. *Raising Less Corn and More Hell*. Urbana and Chicago: University of Illinois Press, 1988.

Tindall, George Brown. *America: A Narrative History*. Vol. 2. New York: W.W. Norton & Company, 1984.

Trawney, R.H. *Religion and the Rise of Capitalism*. New York: New American Library, 1954.

Turner, Frederick Jackson. "The Significance of the Frontier in American History." New York: Frederick Unger Publishing Co., 1979.

Union of Concerned Scientists. "The Gene Exchange: Bt Crops." www.ucsusa.org/Gene/W96.bt.html#bollworm.

U. S. Department of Agriculture. "1997 Census of Agriculture." www.nass.usda.gov/census/census97/highlights/usaum/us.txt

__________________. "Farms and Ranches Decline in 2000." usda.mannlib.cornell.edu/reports/nassr/other/zfl-bb/fmno0201.txt.

U.S. Department of Energy Office of Science, Office of Biological ad Environmental Research, Human Genone Program. "Gene Therapy." www.ornl.gov/hgmis/medicine/genetherapy.html.

Virginia Cooperative Extension. "Boom or Bust for Bt Crops?" filebox.vt.edu/cals/cseschangedor/boombust.html.

Weaver, Richard. *Ideas Have Consequences*. Chicago: University of Chicago Press. 1948.

Wing, Steve and Susanne Wolf. "Intensive Livestock Opera tions, Health and Quality of Life Among Eastern North Carolina Residents."Chapel Hill: North Carolina Department of Health and Human Services, 1999.

RECOMMENDED FURTHER READING

These books were not consulted in the writing of this work, but will be of interest to readers of *The Triumph of Technique.*

Berry, Wendell. *The Unsettling of America.* San Francisco: Sierra Club Books, 1969.

Jackson, Dana L. and Laura L. Jackson. *The Farm as Natural Habitat.* Washington, D.C.: Island Press, 2002.

Kimball, Andrew, ed. *Fatal Harvest: The Tragedy of Industrial Agriculture.* Washington, D.C., 2002.